VOLUME ONE HUNDRED AND THIRTY SEVEN

# Advances in
# Cancer Research

Edited by

**KENNETH D. TEW**

*Medical University of South Carolina,
Charleston, SC, United States*

**PAUL B. FISHER**

*VCU Institute of Molecular Medicine;
VCU Massey Cancer Center,
Virginia Commonwealth University,
School of Medicine,
Richmond, VA, United States*

ACADEMIC

An imprint of Elsevier

Academic Press is an imprint of Elsevier
50 Hampshire Street, 5th Floor, Cambridge, MA 02139, United States
525 B Street, Suite 1800, San Diego, CA 92101-4495, United States
The Boulevard, Langford Lane, Kidlington, Oxford OX5 1GB, United Kingdom
125 London Wall, London, EC2Y 5AS, United Kingdom

First edition 2018

Copyright © 2018 Elsevier Inc. All rights reserved.

No part of this publication may be reproduced or transmitted in any form or by any means, electronic or mechanical, including photocopying, recording, or any information storage and retrieval system, without permission in writing from the publisher. Details on how to seek permission, further information about the Publisher's permissions policies and our arrangements with organizations such as the Copyright Clearance Center and the Copyright Licensing Agency, can be found at our website: www.elsevier.com/permissions.

This book and the individual contributions contained in it are protected under copyright by the Publisher (other than as may be noted herein).

**Notices**
Knowledge and best practice in this field are constantly changing. As new research and experience broaden our understanding, changes in research methods, professional practices, or medical treatment may become necessary.

Practitioners and researchers must always rely on their own experience and knowledge in evaluating and using any information, methods, compounds, or experiments described herein. In using such information or methods they should be mindful of their own safety and the safety of others, including parties for whom they have a professional responsibility.

To the fullest extent of the law, neither the Publisher nor the authors, contributors, or editors, assume any liability for any injury and/or damage to persons or property as a matter of products liability, negligence or otherwise, or from any use or operation of any methods, products, instructions, or ideas contained in the material herein.

ISBN: 978-0-12-815123-5
ISSN: 0065-230X

For information on all Academic Press publications
visit our website at https://www.elsevier.com/books-and-journals

*Publisher:* Zoe Kruze
*Acquisition Editor:* Zoe Kruze
*Editorial Project Manager:* Fenton Coulthurst
*Production Project Manager:* James Selvam
*Cover Designer:* Alan Studholme

Typeset by SPi Global, India

# CONTENTS

Contributors   vii

1. **Unconventional Approaches to Modulating the Immunogenicity of Tumor Cells**   1
   Laurence Booth, Jane L. Roberts, John Kirkwood, Andrew Poklepovic, and Paul Dent

   1. Text Elements   2
   Acknowledgments   11
   References   11
   Further Reading   15

2. **A Theoretical Basis for the Efficacy of Cancer Immunotherapy and Immunogenic Tumor Dormancy: The Adaptation Model of Immunity**   17
   Masoud H. Manjili

   1. Introduction   18
   2. Outcome of Cancer Immunotherapies Inspired by the SNS and Danger Models   19
   3. Beyond the SNS and Danger Models: Tumor Escape and Immune Evasion   22
   4. Discovery and Modulation of Tumor Adaptation Receptors   24
   5. Immunogenic Dormancy of Occult Tumor Cells Through Adaptation   29
   Acknowledgments   31
   Financial Support   31
   Disclosures   31
   References   32

3. **Bcl-2 Antiapoptotic Family Proteins and Chemoresistance in Cancer**   37
   Santanu Maji, Sanjay Panda, Sabindra K. Samal, Omprakash Shriwas, Rachna Rath, Maurizio Pellecchia, Luni Emdad, Swadesh K. Das, Paul B. Fisher, and Rupesh Dash

   1. Introduction   38
   2. Bcl-2 Family Proteins and Chemoresistance   49
   3. Potential Bcl-2 Inhibitors in Chemoresistance: Clinical Development   55
   4. Conclusions and Future Perspectives   63
   Acknowledgments   63
   References   64

4. **New Insights Into Beclin-1: Evolution and Pan-Malignancy Inhibitor Activity** — 77
Stephen L. Wechman, Anjan K. Pradhan, Rob DeSalle, Swadesh K. Das, Luni Emdad, Devanand Sarkar, and Paul B. Fisher

  1. General Introduction — 78
  2. Introduction to Beclin-1 — 79
  3. Beclin-1 and Cancer — 85
  4. Conclusions and Future Directions — 103
  Acknowledgments — 103
  Supplementary Material — 104
  References — 104

5. **Recent Advances in Nanoparticle-Based Cancer Drug and Gene Delivery** — 115
Narsireddy Amreddy, Anish Babu, Ranganayaki Muralidharan, Janani Panneerselvam, Akhil Srivastava, Rebaz Ahmed, Meghna Mehta, Anupama Munshi, and Rajagopal Ramesh

  1. Introduction — 116
  2. Types of Nanoparticles for Therapeutic Delivery — 117
  3. Anticancer Therapeutics and Nanoparticle-Based Delivery Agents — 122
  4. Nanoparticle-Based Codelivery of Drugs and Genes — 136
  5. Stimuli-Responsive Drug Delivery — 139
  6. Nanoparticle-Based Receptor-Targeted Delivery — 148
  7. Conclusion and Prospects — 156
  Acknowledgments — 157
  References — 157
  Further Reading — 170

6. **Evaluation of Resveratrol in Cancer Patients and Experimental Models** — 171
Monica A. Valentovic

  1. Introduction — 172
  2. Resveratrol and Dietary Sources — 172
  3. Resveratrol and Clinical Studies in Cancer Patients — 173
  4. RES Mechanisms for Protection — 177
  5. Nephroprotection by Resveratrol for Cisplatin — 179
  6. Summary — 183
  References — 184

**Monica A. Valentovic**
Toxicology Research Cluster, Joan C. Edward School of Medicine, Huntington, WV, United States

**Stephen L. Wechman**
Virginia Commonwealth University, School of Medicine, Richmond, VA, United States

# CONTRIBUTORS

**Rebaz Ahmed**
Graduate Program in Biomedical Sciences, The University of Oklahoma Health Sciences Center, Oklahoma City, OK, United States

**Narsireddy Amreddy**
Stephenson Cancer Center, The University of Oklahoma Health Sciences Center, Oklahoma City, OK, United States

**Anish Babu**
Stephenson Cancer Center, The University of Oklahoma Health Sciences Center, Oklahoma City, OK, United States

**Laurence Booth**
Virginia Commonwealth University, Richmond, VA, United States

**Swadesh K. Das**
VCU Institute of Molecular Medicine; VCU Massey Cancer Center, Virginia Commonwealth University, School of Medicine, Richmond, VA, United States

**Rupesh Dash**
Institute of Life Sciences, Bhubaneswar, Odisha, India

**Paul Dent**
Virginia Commonwealth University, Richmond, VA, United States

**Rob DeSalle**
Sackler Institute for Comparative Genomics, American Museum of Natural History, New York, NY, United States

**Luni Emdad**
VCU Institute of Molecular Medicine; VCU Massey Cancer Center, Virginia Commonwealth University, School of Medicine, Richmond, VA, United States

**Paul B. Fisher**
VCU Institute of Molecular Medicine; VCU Massey Cancer Center, Virginia Commonwealth University, School of Medicine, Richmond, VA, United States

**John Kirkwood**
University of Pittsburgh Cancer Institute Melanoma and Skin Cancer Program, Hillman Cancer Research Pavilion Laboratory, Pittsburgh, PA, United States

**Santanu Maji**
Institute of Life Sciences, Bhubaneswar, Odisha; Manipal University, Manipal, Karnataka, India

**Masoud H. Manjili**
VCU Institute of Molecular Medicine; VCU Massey Cancer Center, Virginia Commonwealth University, School of Medicine, Richmond, VA, United States

**Meghna Mehta**
Stephenson Cancer Center, The University of Oklahoma Health Sciences Center, Oklahoma City, OK, United States

**Anupama Munshi**
Stephenson Cancer Center, The University of Oklahoma Health Sciences Center, Oklahoma City, OK, United States

**Ranganayaki Muralidharan**
Stephenson Cancer Center, The University of Oklahoma Health Sciences Center, Oklahoma City, OK, United States

**Sanjay Panda**
HCG Panda Cancer Centre; Acharya Harihar Regional Cancer Centre, Cuttack, Odisha, India

**Janani Panneerselvam**
Stephenson Cancer Center, The University of Oklahoma Health Sciences Center, Oklahoma City, OK, United States

**Maurizio Pellecchia**
University of California, Riverside, Riverside, CA, United States

**Andrew Poklepovic**
Virginia Commonwealth University, Richmond, VA, United States

**Anjan K. Pradhan**
Virginia Commonwealth University, School of Medicine, Richmond, VA, United States

**Rajagopal Ramesh**
Stephenson Cancer Center; Graduate Program in Biomedical Sciences, The University of Oklahoma Health Sciences Center, Oklahoma City, OK, United States

**Rachna Rath**
Sriram Chandra Bhanj Dental College and Hospital, Cuttack, Odisha, India

**Jane L. Roberts**
Virginia Commonwealth University, Richmond, VA, United States

**Sabindra K. Samal**
Institute of Life Sciences, Bhubaneswar, Odisha; Manipal University, Manipal, Karnataka, India

**Devanand Sarkar**
VCU Institute of Molecular Medicine; VCU Massey Cancer Center, Virginia Commonwealth University, School of Medicine, Richmond, VA, United States

**Omprakash Shriwas**
Institute of Life Sciences, Bhubaneswar, Odisha; Manipal University, Manipal, Karnataka, India

**Akhil Srivastava**
Stephenson Cancer Center, The University of Oklahoma Health Sciences Center, Oklahoma City, OK, United States

CHAPTER ONE

# Unconventional Approaches to Modulating the Immunogenicity of Tumor Cells

**Laurence Booth\*, Jane L. Roberts\*, John Kirkwood[†], Andrew Poklepovic\*, Paul Dent\*,[1]**

\*Virginia Commonwealth University, Richmond, VA, United States
[†]University of Pittsburgh Cancer Institute Melanoma and Skin Cancer Program, Hillman Cancer Research Pavilion Laboratory, Pittsburgh, PA, United States
[1]Corresponding author: e-mail address: paul.dent@vcuhealth.org

## Contents

| | |
|---|---|
| 1. Text Elements | 2 |
| Acknowledgments | 11 |
| References | 11 |
| Further Reading | 15 |

## Abstract

For several years, it has been known that histone deacetylase inhibitors have the potential to alter the immunogenicity of tumor cells exposed to checkpoint inhibitory immunotherapy antibodies. HDAC inhibitors can rapidly reduce expression of PD-L1 and increase expression of MHCA in various tumor types that subsequently facilitate the antitumor actions of checkpoint inhibitors. Recently, we have discovered that drug combinations which cause a rapid and intense autophagosome formation also can modulate the expression of HDAC proteins that control tumor cell immunogenicity via their regulation of PD-L1 and MHCA. These drug combinations, in particular those using the irreversible ERBB1/2/4 inhibitor neratinib, can result in parallel in the internalization of growth factor receptors as well as fellow-traveler proteins such as mutant K-RAS and mutant N-RAS into autophagosomes. The drug-induced autophagosomes contain HDAC proteins/signaling proteins whose expression is subsequently reduced by lysosomal degradation processes. These findings argue that cancer therapies which strongly promote autophagosome formation and autophagic flux may facilitate the subsequent use of additional antitumor modalities using checkpoint inhibitor antibodies.

## 1. TEXT ELEMENTS

Histone deacetylase inhibitors have been under investigation as anticancer agents for over 20 years (Zhan, Wang, Liu, & Suzuki, 2017). Simplistically, HDAC inhibitors regulate the acetylation status of histones, proteins that in turn regulate the condensation status of DNA, and the accessibility of promoter and suppressor elements to transcription factors, thereby regulating transcription. However, multiple other cytosolic and nuclear proteins are also regulated by reversible acetylation. Two of the most notable acetylated proteins whose functions are of prime importance in the survival of many tumor cell types are heat shock protein 90 (HSP90) and the p65 subunit of NFκB (Leus, Zwinderman, & Dekker, 2016; Rodrigues, Thota, & Fraga, 2016).

Acetylation of p65 NFκB plays a key role in activation of the transcription factor. For drugs that utilize NFκB signaling as a component of their "cell death signal," e.g., by elevating TNFα expression, HDAC inhibitors will facilitate p65 acetylation and tumor cell killing (Gang, Shaw, Dhingra, Davie, & Kirshenbaum, 2013). However, for drugs that use compensatory NFκB activation to protect themselves from a toxic stress, HDAC inhibitors have the potential via NFκB to suppress cell death (Karthik, Sankar, Varunkumar, Anusha, & Ravikumar, 2015). As single agents at clinically relevant concentrations, HDAC inhibitors often cause modest levels to tumor cell killing; the combination of HDAC inhibitors with agents that block NFκB activation, however, results in a synergy of tumor cell killing (Li, Li, et al., 2016; Li, Zhuang, et al., 2016). Multiple other transcription factors are regulated by reversible acetylation including p53, STAT3, GATA-1, and Sp3 (Formisano et al., 2015; Schäfer et al., 2017; Watamoto et al., 2003; Yuan, Guan, Chatterjee, & Chin, 2005). HSP90 acetylation is regulated by the enzyme HDAC6 and the acetyltransferase that also associates with HSP90, arrest defective-1 protein (ARD1) (DePaolo et al., 2016; Yang, Zhang, Zhang, Zhang, & Xu, 2013). Hyperacetylation of HSP90 has been proposed to cause the release of the cochaperone complex protein p23, and to inhibit the chaperone's ATPase function, collectively reducing HSP90 chaperoning activity (Bali, Pranpat, Bradner, et al., 2005; Kekatpure, Dannenberg, & Subbaramaiah, 2009; Koga et al., 2006; Rao et al., 2008). Other chaperone proteins, e.g., HSP70 and GRP78 have also been found to be regulated by reversible acetylation (Chang et al., 2016; Li, Li, et al., 2016; Li, Zhuang, et al., 2016; Park, Seo, Park, Lee, & Kim, 2017; Seo et al., 2016). Acetylation of HSP90 has been proposed to regulate it and its client proteins ubiquitination and subsequent proteolytic breakdown (Mollapour & Neckers, 2012; Nanduri, Hao, Fitzpatrick, & Yao, 2015; Quadroni, Potts, & Waridel, 2015; Zhou, Agoston, Atadja, Nelson, & Davidson, 2008).

Immunotherapy, using checkpoint inhibitory antibodies, has become a first line therapeutic regimen in melanoma, NSCLC, bladder cancer, and H&N SCC. Antibodies that blockade the functions of PD-1, PD-L1, and CTLA-4 have all been approved as therapeutics within the last 5 years (Emens et al., 2017; Koller et al., 2016). Histone deacetylase inhibitors are known to increase MHC class I and II expression on the cell surface which would facilitate antitumor responses from both the innate and the adaptive immune systems (Nakajima et al., 2017). HDAC inhibitors have been shown to activate NK cells (Tiper & Webb, 2016). Other studies have linked HDAC inhibitors to both increased and decreased expression of PD-L1 and PD-L2 on tumor cells with the differential effects appearing to be dependent on HDAC inhibitor dose or the cell lines being tested, though all studies argue that HDAC inhibitors enhance the antitumor responses of the immune system using checkpoint inhibitory antibodies (Beg & Gray, 2016; Shen, Orillion, & Pili, 2016; Terranova-Barberio, Thomas, & Munster, 2016; Yang et al., 2015; Zheng et al., 2016). Thus, HDAC inhibitors have the potential to enhance the efficacy of checkpoint inhibitory antibodies. And, several pan-HDAC inhibitors, such as sodium valproate, etinostat, panobinostat, and vorinostat, have all been shown in vivo to enhance the antitumor efficacy of checkpoint inhibitory antibodies, as well as promote activated T cell infiltration into the tumors (Booth, Roberts, Poklepovic, & Dent, 2017; Booth, Roberts, Poklepovic, Kirkwood, & Dent, 2017; Christiansen et al., 2011; Gameiro, Malamas, Tsang, Ferrone, & Hodge, 2016; Hornig, Heppt, Graf, Ruzicka, & Berking, 2016; Kroesen et al., 2016; Vo et al., 2009; West & Johnstone, 2014; West et al., 2013). Tumor types tested in these studies are diverse and include melanoma, breast, colorectal, glioblastoma, and hepatoma (Chae, Wang, Nimeiri, Kalyan, & Giles, 2017; Huang et al., 2017; Keller, Zhang, Li, Schaider, & Wells, 2017; Monnot & Romero, 2017; Rai et al., 2017).

There are several FDA-approved HDAC inhibitors and also FDA-approved anti-PD-1 antibodies (pembrolizumab and nivolumab). And, at present there are a number of clinical trials are evaluating the HDAC inhibitor + anti-PD-1 antibody combination in a variety of tumor types. One phase II trial will use the anti-PD-1 antibody pembrolizumab in combination with pan-HDAC inhibitor vorinostat (NCT02638090). Pembrolizumab, nivolumab (anti-PD-1), and atezolizumab (anti-PD-L1) antibodies are being combined with the class I specific HDAC inhibitor entinostat in several clinical trials, respectively (NCT02437136, NCT02697630, NCT01928576, NCT02453620, and NCT02708680). At present, there is an open phase I clinical trial in which the stated HDAC6-specific inhibitor rocilinostat (ACY-241) is being combined with anti-PD-1 and anti-CTLA4 antibodies in malignant melanoma (NCT02935790). The trial NCT01928576 will determine the safety and efficacy of entinostat and azacytidine, a DNA methyl-transferase inhibitor, on the response to nivolumab.

Studies by our laboratory in the field of immunotherapy were initiated in the autumn of 2016 using the pan-HDACi drugs sodium valproate and AR42, and using B16 melanoma cells as an in vivo model system (Booth, Roberts, Poklepovic, Kirkwood, & Dent, 2017; Booth, Roberts, Sander, et al., 2017). Our in vivo studies were predicated on in vitro findings which demonstrated valproate and AR42 both rapidly reduced the expression of PD-L1 and ornithine decarboxylase (ODC) in multiple tumor cell types, and enhanced the expression of class I MHCA. These effects on protein expression were enhanced when the HDAC inhibitor was combined with the multikinase and chaperone inhibitor pazopanib, which together also facilitated the release into the extracellular environment of the immunogenic nuclear protein HMGB1. As presented in Fig. 1, a 35-day prior treatment of B16 melanoma tumors with the pan-HDAC inhibitor AR42 results in the infiltration of tumors with M1-polarized macrophages, which predicts for a strong antitumor immune response. See Fig. 2 for the mechanisms by

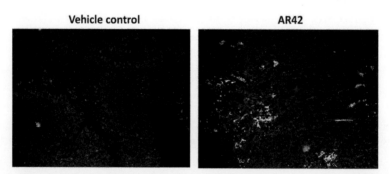

**Fig. 1** The pan-HDAC inhibitor AR42 enhances M1 macrophage infiltration into the tumor 5 weeks after drug exposure. Trametinib/dabrafenib-resistant MEL28 melanoma cells were implanted into the rear flanks of athymic mice. Tumors were treated with vehicle control or with AR42 (15 mg/kg) for 2 days. Thirty-five days later, at the time of animal nadir, tumors were fixed, embedded in paraffin, and 4-μm sections obtained. Sections were deparaffinized, renatured, and stained to examine (10× mag.) the colocalization of F4/80 staining and iNOS staining (that stains for M1 macrophages).

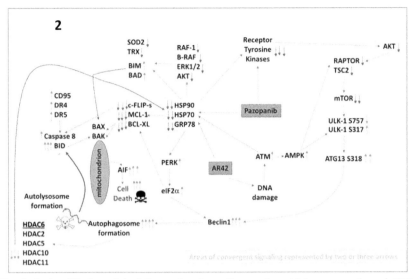

**Fig. 2** A simplified model of the molecular pathways by which pazopanib and AR42 combine to kill cancer cells. As an HDAC inhibitor, AR42 causes DNA damage and causes inhibitory chaperone acetylation. Pazopanib, as a multikinase and chaperone inhibitor, also inhibits chaperone activities as well as many class III receptor tyrosine kinases. DNA damage causes activation of ATM. ATM signals to activate the AMPK. AMPK signaling inactivates RAPTOR and TSC2 resulting in the inactivation of mTORC1 and mTORC2. Downstream of mTOR is the kinase ULK-1; the drug combination via AMPK promotes ULK-1 S317 phosphorylation which activates the kinase; the drug combination via mTOR inactivation reduces ULK-1 S757 phosphorylation which also activates the kinase. Activated ULK-1 phosphorylates ATG13 which is the key gate-keeper step in permitting autophagosome formation. AR42-induced ATM signaling also acts to reduce the activities of multiple chaperone proteins. Reduced HSP90 and HSP70 function lowers the expression of all receptor tyrosine kinases and the activities of STAT3, STAT5, ERK1/2, and AKT that results in lower expression of ROS/RNS detoxifying enzymes such as TRX and SOD2. Reduced GRP78 function causes activation of PERK and subsequently eIF2α. Enhanced eIF2α signaling reduces the transcription of proteins with short half-lives such as c-FLIP-s, MCL-1, and BCL-XL, and enhances expression of Beclin1, DR4, and DR5. Thus, the convergent actions of reduced HSP90 and HSP70 chaperone activity and eIF2α signaling lead to a profound reduction in the protein levels of c-FLIP-s, MCL-1, and BCL-XL which facilitates death receptor signaling through CD95, DR4, and DR5 to activate the extrinsic apoptosis pathway. Enhanced Beclin1 expression converges with elevated ATG13 phosphorylation to produce high levels of autophagosome formation that acts to reduce HDAC2/5/6/10/11 expression but also to stall autophagosome fusion with lysosomes and to stall autolysosome maturation, which likely through cytosolic cathepsin proteases converges with the extrinsic apoptosis pathway to cleave BID and cause mitochondrial dysfunction. Tumor cell killing downstream of the mitochondrion was mediated by AIF and not caspases 3/7. The tumoricidal actions of AIF were facilitated by reduced HSP70 functionality as this chaperone can sequester AIF in the cytosol and prevent its translocation to the nucleus.

which pazopanib and HDAC inhibitors interact to kill melanoma cells. Both valproate and AR42 were shown to enhance the anti-B16 tumor efficacy of an anti-PD-1 antibody and of an anti-CTLA4 antibody.

In contemporaneous studies, we had discovered that the drug combination of [pemetrexed+sildenafil] reduced the expression of HDAC6 which resulted in elevated HSP90 acetylation (Booth, Roberts, Poklepovic, Gordon, & Dent, 2017). Acetylated HSP90 exhibited lower ATPase activity and had reduced association with client proteins. Although HDAC6 is known to associate with components of the ubiquitin/proteasome system and be ubiquitinated itself, the rapid reduction in HDAC6 expression was prevented by knock down of Beclin1 or ATG5, essential proteins in the regulation of autophagosome formation. The proteasome inhibitor bortezomib did not prevent the degradation of HDAC6. These findings raised the possibility that autophagosome formation could regulate the acetylation of proteins, indirectly, through modulation of HDAC protein levels.

Thus, additional studies were performed to determine whether the [pazopanib+HDAC inhibitor] combination or the previously established drug combinations of [pemetrexed+sorafenib] or [pemetrexed+sildenafil], that we have shown to also kill through autophagy, regulated the expression of not only HDAC6 but also of HDACs1–11 (Booth, Roberts, Poklepovic, Gordon, et al., 2017; Booth, Roberts, Poklepovic, Kirkwood, & Dent, 2017; Booth et al., 2016). It was discovered that [pemetrexed+sildenafil], [pazopanib+HDAC inhibitor], and [pemetrexed+sorafenib] all acted to rapidly, within 6h, reduce the expression of multiple HDAC proteins, and effect blocked by knock down of Beclin1 or of ATG5. See Fig. 3 for the mechanisms by which pemetrexed, sorafenib, and neratinib interact to kill tumor cells. Again, use of bortezomib to block proteasome function did not prevent HDAC degradation. In tumor types tested, including melanoma, sarcoma, lung and ovarian, [pemetrexed+sildenafil], and [pemetrexed+sorafenib] routinely reduced the expression of HDAC2, HDAC4, HDAC6, HDAC9, and HDAC11, respectively. Treatment of tumor cells with [pazopanib+HDAC inhibitor] routinely reduced the levels of HDAC1, HDAC2, HDAC3, HDAC5, HDAC6, HDAC7, HDAC9, HDAC10, and HDAC11. Based on the tumor cell, the HDACs responsible for the regulation/inhibition of promoter elements for PD-L1, MHCA, and ODC are reported to be HDAC1, HDAC2, HDAC3, and HDAC10. Molecular studies, knocking down the expression of individual HDAC proteins or combinations of HDAC proteins revealed that PD-L1 expression was coordinately regulated by HDACs1/2/3/10. The expression of MHCA was most often regulated by HDACs1/3/10 and the expression of ODC regulated by

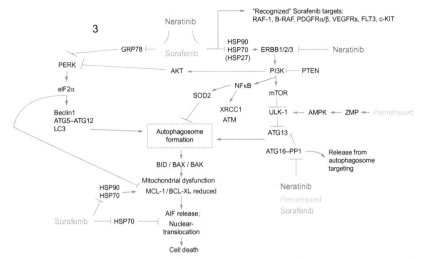

**Fig. 3** A simplified model of the molecular pathways by which pemetrexed, sorafenib, and neratinib combine to kill cancer cells. As a thymidylate synthase inhibitor, pemetrexed causes DNA damage and elevates ZMP levels. ATM phosphorylates the AMPK and ZMP allosterically activates the AMPK which leads to inactivation of mTOR, activation of ULK-1 and ATG13 phosphorylation concomitant with formation of toxic autophagosomes. Sorafenib, in addition to inhibiting RAF kinases and class III RTKs, is an HSP90 family and HSP70 family chaperone inhibitor. Reduced GRP78 function facilitates and endoplasmic reticulum stress response which acts to enhance the levels of Beclin1 and ATG5, facilitating autophagosome formation. The induction of autophagy results in mitochondrial dysfunction which is facilitated by ER stress acting to downregulate MCL-1 and BCL-XL expression. Sorafenib, via inhibition of HSP90 and HSP70, permits AIF, released from the mitochondria, to dissociate from inhibitory chaperones and where it then translocates to the nucleus.

HDACs2/3/10. Of note was that the expression of PD-L1, MHCA, and ODC was more weakly modulated by knock down of HDAC6, compared to HDACs1/2/3/10, arguing that the changes in PD-L1, MHCA, and ODC expression were primarily due to altered transcription and not protein stability.

In vivo studies using HDAC inhibitors alone or in combination with anti-PD-1 or anti-CTLA4 antibodies demonstrated that a transient exposure of established B16 melanoma tumors to HDAC inhibitors resulted in a permanent (~20 days) upregulation of MHCA expression and downregulation of PD-L1 levels within the tumor cells (Booth, Roberts, Poklepovic, Kirkwood, Dent, 2017). These observations in drug-treated tumors correlated with increased infiltration into the tumors of M1 macrophages, NK cells, and activated T cells (see Fig. 1). The presence of M1 macrophages, NK cells, and activated T cells would all be predicted to act in an antitumor fashion to suppress tumor growth. The HDAC inhibitors sodium valproate and AR42 each enhanced the antitumor efficacy of an anti-PD-1 antibody and of an anti-CTLA4 antibody. Other in vivo studies treated Lewis lung carcinoma tumors with [pemetrexed + sildenafil], and determined that this drug combination also enhanced the antitumor efficacy of an anti-PD-1 antibody and of an anti-CTLA4 antibody (Booth, Roberts, Poklepovic, & Dent, 2017). Collectively, these findings all confirm that HDAC inhibitors, and combinations of agents that act to reduce HDAC levels, can enhance the efficacy of checkpoint immunotherapeutic antibodies in vivo.

More recent studies from our laboratory have taken the concept of autophagic degradation of HDAC proteins and other signaling mediators to a new level of complexity. Treatment of tumor cells with clinically relevant concentrations of the irreversible ERBB1/2/4 inhibitor neratinib (100 nM), and to a greater extent [neratinib + sodium valproate] also can reduce the expression of HDACs via autophagic degradation (Booth, Roberts, Poklepovic, Avogadri-Connors, et al., 2017). In turn, this results in reduced PD-L1, PD-L2, and ODC expression, and increased MHCA levels. We discovered that [neratinib + sodium valproate] enhanced the antitumor activity of an anti-PD-1 antibody using the 4T1 mouse mammary carcinoma model system. Thus, we predict that many drug combination therapies that promote autophagosome formation and autophagic flux could act to reduce HDAC expression, and thus regulate the protein levels of PD-L1, MHCA, and ODC, and tumor cell immunogenicity.

During our neratinib studies we became aware that neratinib, acting solely as an irreversible ERBB1/2/4 inhibitor, was only one component of a complicated biology for this drug (Booth, Roberts, Poklepovic, Avogadri-Connors, et al., 2017; Booth, Roberts, Poklepovic, Kirkwood, Sander, et al., 2017). Initial studies with neratinib were designed to examine the effects of neratinib on ERBB1 phosphorylation in NSCLC cells expressing a mutated active ERBB1 protein. Although neratinib significantly reduced ERBB1 phosphorylation, it also rapidly, within 6 h, reduced total ERBB1 expression. In mammary and ovarian cancer cells expressing ERBB2, a similar phenomenon occurred when cells were treated with neratinib. Combined incubation of neratinib-treated cells with either sodium valproate or the novel HDAC inhibitor AR42 enhanced the downregulation of ERBB1 and ERBB2 expression. Knock down of either Beclin1 or ATG5 prevented neratinib or [neratinib + HDAC inhibitor] from reducing ERBB1 and ERBB2 expression.

As negative controls in our neratinib studies, we included assessments of changes in ERBB3 and c-MET expression. ERBB3 lacks an active catalytic site and in theory should not bind neratinib, and c-MET, which is not part of the ERBB receptor family, should not bind neratinib. To our surprise, in addition to ERBB1 and ERBB2, neratinib could also reduce both ERBB3 and c-MET expression. The time course of ERBB3 and c-MET down-regulation appeared to be slightly slower than that observed for ERBB1 and ERBB2 suggesting their degradation were secondary processes. Knock down of Beclin1 or ATG5 prevented the downregulation of c-MET. Follow up findings also demonstrated that the levels of PDGFRα could be reduced by neratinib in GBM cells (Booth, Roberts, Poklepovic, Kirkwood, Sander, et al., 2017). See Fig. 4 for the mechanisms by which neratinib and sodium valproate interact to kill tumor cells.

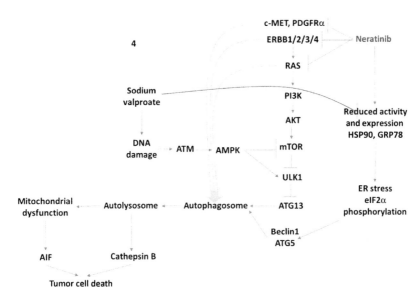

**Fig. 4** A simplified model of the molecular pathways by which neratinib and valproate combine to kill cancer cells. As an HDAC inhibitor, sodium valproate causes DNA damage. This activates an ATM–AMPK pathway that inactivates mTOR, activated ULK1, and results in ATG13 phosphorylation and autophagosome formation. Neratinib both inhibits and downregulates the expression of ERBB family receptors as well as other fellow-traveler RTKs and mutant RAS proteins. Sodium valproate enhances HSP90 and GRP78 acetylation which reduces chaperone activity and neratinib reduces the protein levels of HSP90 and GRP78; this results in an endoplasmic reticulum stress response, phosphorylation of eIF2α, and increased expression of Beclin1 and ATG5, that in turn facilitate autophagosome formation. Cathepsin B, from autolysosomes both cleaves BID to cause mitochondrial dysfunction and acts as a direct cell killing protease. Elevated ER stress signaling reduces MCL-1 and BCL-XL levels which facilitates AIF release and AIF translocation to the nucleus where it executes the tumor cell.

The logical progression from these discoveries was to examine the impact of neratinib on the levels of other plasma membrane-associated proteins which can play a tumor-promoting role. Mutated RAS proteins have been considered for several decades to be one of the most important cancer-driving proteins to therapeutically target (McCormick, 2015). RAS proteins, to be active and plasma membrane localized, are prenylated, and this essential modification can be targeted using specific farnesyl transferase inhibitors or inhibitors of HMG CoA reductase (i.e., statins) (Hamed et al., 2008). However, K-RAS and N-RAS proteins, in addition to farnesylation, can also be alternatively geranyl geranylated. As a result, farnesyl transferase inhibitors have largely been unsuccessful in the clinic at controlling RAS-dependent tumors. In our most recent studies, we demonstrated that clinically relevant concentrations of neratinib and [neratinib + valproate] activate the AMPK which in turn phosphorylates and inactivates HMG Co-A reductase (Booth, Roberts, Poklepovic, Kirkwood, Sander, et al., 2017). Reduced HMG Co-A reductase activity will lower the levels of mevalonate that ultimately results in reduced levels of both the farnesyl and geranyl substrates. Thus, the use of the [neratinib + valproate] combination attacks the activity of mutant RAS proteins by direct and indirect mechanisms. The drug combination directly reduces RAS protein levels through a process requiring autophagy and lysosomal degradation, and the combination indirectly reduces RAS function by impeding the ability of the cell to prenylate and facilitate plasma membrane localization of RAS proteins. Thus, our data argue that the degradation of other receptors and RAS proteins by neratinib is a secondary on-target effect of the drug and occurs because of the primary inhibition, internalization, and degradation of ERBB family receptors.

This review began by discussing how the modulation of HDAC function, via HDAC inhibitors or via HDAC protein degradation, could regulate the expression of PD-L1 and MHCA, and tumor cell immunogenicity. As our studies led us toward a more complicated understanding of the mechanisms of action for neratinib, we realized that neratinib alone, or in combination with HDAC inhibitors, could suppress signaling from multiple growth factor receptors and oncogenic mutant RAS proteins. Thus, does receptor-RAS signaling play a role in the protein levels of PD-L1 and MHCA? In NSCLC cells, and other tumor types, e.g., HNSCC, breast cancer, signaling by activated ERBB1 or mutant K-RAS, via the ERK1/2 pathway, acts to increase the expression of PD-L1 and MHCA (El-Jawhari et al., 2014; Ji et al., 2016; Loi et al., 2016; Sers et al., 2009; Sumimoto, Takano, Teramoto, & Daigo,

2016; Yang et al., 2017; Zhang et al., 2017). Inhibitors of ERBB1 or of MEK1/2 suppress the abilities of mutant ERBB1 or mutant RAS from maintaining PD-L1 and MHCA expression. Thus, neratinib and valproate exposure act to enhance tumor cell immunogenicity both via reducing HDAC expression and derepressing ERK1/2 signaling that prevents upregulation of immunotherapy biomarkers.

In conclusion, we have demonstrated that multiple drug combinations that promote autophagosome formation all have the theoretical potential to downregulate the expression of HDAC proteins thereby opsonizing tumor cells to checkpoint immunotherapies. By also reducing signaling through the ERK1/2 pathway, many of these drug combinations also derepress gene expression elements facilitating increases in MHCA and decreases in PD-L1.

## ACKNOWLEDGMENTS

Support for the present study was funded from philanthropic funding from Massey Cancer Center, the Universal Inc. Chair in Signal Transduction Research and PHS R01-CA192613.

## REFERENCES

Bali, P., Pranpat, M., Bradner, J., et al. (2005). Inhibition of histone deacetylase 6 acetylates and disrupts the chaperone function of heat shock protein 90—A novel basis for antileukemia activity of histone deacetylase inhibitors. *The Journal of Biological Chemistry*, *280*, 26729–26734.

Beg, A. A., & Gray, J. E. (2016). HDAC inhibitors with PD-1 blockade: A promising strategy for treatment of multiple cancer types? *Epigenomics*, *8*, 1015–1017.

Booth, L., Roberts, J. L., Poklepovic, A., Avogadri-Connors, F., Cutler, R. E., Lalani, A. S., et al. (2017). HDAC inhibitors enhance neratinib activity and when combined enhance the actions of an anti-PD-1 immunomodulatory antibody in vivo. *Oncotarget*, *8*(52), 90262–90277.

Booth, L., Roberts, J. L., Poklepovic, A., & Dent, P. (2017). [Pemetrexed + sildenafil] regulates the immunotherapy response of tumor cells. *Cancer Biology and Therapy*, *18*(9), 705–714.

Booth, L., Roberts, J. L., Poklepovic, A., Gordon, S., & Dent, P. (2017). PDE5 inhibitors enhance the lethality of pemetrexed through inhibition of multiple chaperone proteins and via the actions of cyclic GMP and nitric oxide. *Oncotarget*, *8*, 1449–1468.

Booth, L., Roberts, J. L., Poklepovic, A., Kirkwood, J., & Dent, P. (2017). HDAC inhibitors enhance the immunotherapy response of melanoma cells. *Oncotarget*, *8*, 83155–83170. https://doi.org/10.18632/oncotarget.17950.

Booth, L., Roberts, J. L., Poklepovic, A., Kirkwood, J., Sander, C., Avogandri-Connors, F., et al. (2017). The levels of mutant K-RAS and mutant N-RAS are rapidly reduced in a Beclin1/ATG5-dependent fashion by the irreversible ERBB1/2/4 inhibitor neratinib. *Cancer Biology & Therapy,* (in press).

Booth, L., Roberts, J. L., Sander, C., Lee, J., Kirkwood, J. M., Poklepovic, A., et al. (2017). The HDAC inhibitor AR42 interacts with pazopanib to kill trametinib/dabrafenib-resistant melanoma cells in vitro and in vivo. *Oncotarget*, *8*, 16367–16386.

Booth, L., Roberts, J. L., Tavallai, M., Chuckalovcak, J., Stringer, D. K., Koromilas, A. E., et al. (2016). [Pemetrexed + Sorafenib] lethality is increased by inhibition of ERBB1/2/3-PI3K-NFκB compensatory survival signaling. *Oncotarget, 7*, 23608–23632.

Chae, Y. K., Wang, S., Nimeiri, H., Kalyan, A., & Giles, F. J. (2017). Pseudoprogression in microsatellite instability-high colorectal cancer during treatment with combination T cell mediated immunotherapy: A case report and literature review. *Oncotarget, 8*, 57889–57897. https://doi.org/10.18632/oncotarget.18361.

Chang, Y. W., Tseng, C. F., Wang, M. Y., Chang, W. C., Lee, C. C., Chen, L. T., et al. (2016). Deacetylation of HSPA5 by HDAC6 leads to GP78-mediated HSPA5 ubiquitination at K447 and suppresses metastasis of breast cancer. *Oncogene, 35*, 1517–1528.

Christiansen, A. J., West, A., Banks, K. M., Haynes, N. M., Teng, M. W., Smyth, M. J., et al. (2011). Eradication of solid tumors using histone deacetylase inhibitors combined with immune-stimulating antibodies. *Proceedings of the National Academy of Sciences of the United States of America, 108*, 4141–4146.

DePaolo, J. S., Wang, Z., Guo, J., Zhang, G., Qian, C., Zhang, H., et al. (2016). Acetylation of androgen receptor by ARD1 promotes dissociation from HSP90 complex and prostate tumorigenesis. *Oncotarget, 7*, 71417–71428.

El-Jawhari, J. J., El-Sherbiny, Y. M., Scott, G. B., Morgan, R. S., Prestwich, R., Bowles, P. A., et al. (2014). Blocking oncogenic RAS enhances tumour cell surface MHC class I expression but does not alter susceptibility to cytotoxic lymphocytes. *Molecular Immunology, 58*, 160–168.

Emens, L. A., Ascierto, P. A., Darcy, P. K., Demaria, S., Eggermont, A. M. M., Redmond, W. L., et al. (2017). Cancer immunotherapy: Opportunities and challenges in the rapidly evolving clinical landscape. *European Journal of Cancer, 81*, 116–129.

Formisano, L., Guida, N., Valsecchi, V., Cantile, M., Cuomo, O., Vinciguerra, A., et al. (2015). Sp3/REST/HDAC1/HDAC2 complex represses and Sp1/HIF-1/p300 complex activates ncx1 gene transcription, in brain ischemia and in ischemic brain preconditioning, by epigenetic mechanism. *The Journal of Neuroscience, 35*, 7332–7348.

Gameiro, S. R., Malamas, A. S., Tsang, K. Y., Ferrone, S., & Hodge, J. W. (2016). Inhibitors of histone deacetylase 1 reverse the immune evasion phenotype to enhance T-cell mediated lysis of prostate and breast carcinoma cells. *Oncotarget, 7*, 7390–7402.

Gang, H., Shaw, J., Dhingra, R., Davie, J. R., & Kirshenbaum, L. A. (2013). Epigenetic regulation of canonical TNFα pathway by HDAC1 determines survival of cardiac myocytes. *American Journal of Physiology. Heart and Circulatory Physiology, 304*, H1662–9.

Hamed, H., Mitchell, C., Park, M. A., Hanna, D., Martin, A. P., Harrison, B., et al. (2008). Human chorionic gonadotropin (hCG) interacts with lovastatin and ionizing radiation to modulate prostate cancer cell viability in vivo. *Cancer Biology & Therapy, 7*, 587–593.

Hornig, E., Heppt, M. V., Graf, S. A., Ruzicka, T., & Berking, C. (2016). Inhibition of histone deacetylases in melanoma—A perspective from bench to bedside. *Experimental Dermatology, 25*, 831–838.

Huang, J., Liu, F., Liu, Z., Tang, H., Wu, H., Gong, Q., et al. (2017). Immune checkpoint in glioblastoma: Promising and challenging. *Frontiers in Pharmacology, 8*, 242.

Ji, M., Liu, Y., Li, Q., Li, X., Ning, Z., Zhao, W., et al. (2016). PD-1/PD-L1 expression in non-small-cell lung cancer and its correlation with EGFR/KRAS mutations. *Cancer Biology & Therapy, 17*, 407–413.

Karthik, S., Sankar, R., Varunkumar, K., Anusha, C., & Ravikumar, V. (2015). Blocking NF-κB sensitizes non-small cell lung cancer cells to histone deacetylase inhibitor induced extrinsic apoptosis through generation of reactive oxygen species. *Biomedicine & Pharmacotherapy, 69*, 337–344.

Kekatpure, V. D., Dannenberg, A. J., & Subbaramaiah, K. (2009). HDAC6 modulates Hsp90 chaperone activity and regulates activation of aryl hydrocarbon receptor signaling. *The Journal of Biological Chemistry, 284*, 7436–7445.

Keller, H. R., Zhang, X., Li, L., Schaider, H., & Wells, J. W. (2017). Overcoming resistance to targeted therapy with immunotherapy and combination therapy for metastatic melanoma. *Oncotarget, 8*, 75675–75686. https://doi.org/10.18632/oncotarget.18523.

Koga, F., Xu, W., Karpova, T. S., McNally, J. G., Baron, R., & Neckers, L. (2006). Hsp90 inhibition transiently activates Src kinase and promotes Src-dependent Akt and Erk activation. *Proceedings of the National Academy of Sciences of the United States of America, 103*, 11318–11322.

Koller, K. M., Wang, W., Schell, T. D., Cozza, E. M., Kokolus, K. M., Neves, R. I., et al. (2016). Malignant melanoma—The cradle of anti-neoplastic immunotherapy. *Critical Reviews in Oncology/Hematology, 106*, 25–54.

Kroesen, M., Büll, C., Gielen, P. R., Brok, I. C., Armandari, I., Wassink, M., et al. (2016). Anti-GD2 mAb and Vorinostat synergize in the treatment of neuroblastoma. *OncoImmunology, 5*, e1164919.

Leus, N. G., Zwinderman, M. R., & Dekker, F. J. (2016). Histone deacetylase 3 (HDAC 3) as emerging drug target in NF-κB-mediated inflammation. *Current Opinion in Chemical Biology, 33*, 160–168.

Li, Z. Y., Li, Q. Z., Chen, L., Chen, B. D., Wang, B., Zhang, X. J., et al. (2016). Histone deacetylase inhibitor RGFP109 overcomes temozolomide resistance by blocking NF-κB-dependent transcription in glioblastoma cell lines. *Neurochemical Research, 41*, 3192–3205.

Li, Z., Zhuang, M., Zhang, L., Zheng, X., Yang, P., & Li, Z. (2016). Acetylation modification regulates GRP78 secretion in colon cancer cells. *Scientific Reports, 6*, 30406.

Loi, S., Dushyanthen, S., Beavis, P. A., Salgado, R., Denkert, C., Savas, P., et al. (2016). RAS/MAPK activation is associated with reduced tumor-infiltrating lymphocytes in triple-negative breast cancer: Therapeutic cooperation between MEK and PD-1/PD-L1 immune checkpoint inhibitors. *Clinical Cancer Research, 22*, 1499–1509.

McCormick, F. (2015). KRAS as a therapeutic target. *Clinical Cancer Research, 21*, 1797–1801.

Mollapour, M., & Neckers, L. (2012). Post-translational modifications of Hsp90 and their contributions to chaperone regulation. *Biochimica et Biophysica Acta, 1823*, 648–655.

Monnot, G. C., & Romero, P. (2017). Rationale for immunological approaches to breast cancer therapy. *Breast*. https://doi.org/10.1016/j.breast.2017.06.009. pii: S0960-9776(17)30477-0.

Nakajima, N. I., Niimi, A., Isono, M., Oike, T., Sato, H., Nakano, T., et al. (2017). Inhibition of the HDAC/Suv39/G9a pathway restores the expression of DNA damage-dependent major histocompatibility complex class I-related chain A and B in cancer cells. *Oncology Reports, 38*, 693–702.

Nanduri, P., Hao, R., Fitzpatrick, T., & Yao, T. P. (2015). Chaperone-mediated 26S proteasome remodeling facilitates free K63 ubiquitin chain production and aggresome clearance. *The Journal of Biological Chemistry, 290*, 9455–9464.

Park, Y. H., Seo, J. H., Park, J. H., Lee, H. S., & Kim, K. W. (2017). Hsp70 acetylation prevents caspase-dependent/independent apoptosis and autophagic cell death in cancer cells. *International Journal of Oncology, 51*, 573–578. https://doi.org/10.3892/ijo.2017.4039.

Quadroni, M., Potts, A., & Waridel, P. (2015). Hsp90 inhibition induces both protein-specific and global changes in the ubiquitinome. *Journal of Proteomics, 120*, 215–229.

Rai, V., Abdo, J., Alsuwaidan, A. N., Agrawal, S., Sharma, P., & Agrawal, D. K. (2017). Cellular and molecular targets for the immunotherapy of hepatocellular carcinoma. *Molecular and Cellular Biochemistry*. https://doi.org/10.1007/s11010-017-3092-z.

Rao, R., Fiskus, W., Yang, Y., Lee, P., Joshi, R., Fernandez, P., et al. (2008). HDAC6 inhibition enhances 17-AAG-mediated abrogation of hsp90 chaperone function in human leukemia cells. *Blood, 112*, 1886–1893.

Rodrigues, D. A., Thota, S., & Fraga, C. A. (2016). Beyond the selective inhibition of histone deacetylase 6. *Mini Reviews in Medicinal Chemistry, 16*, 1175–1184.

Schäfer, C., Göder, A., Beyer, M., Kiweler, N., Mahendrarajah, N., Rauch, A., et al. (2017). Class I histone deacetylases regulate p53/NF-κB crosstalk in cancer cells. *Cellular Signalling, 29*, 218–225.

Seo, J. H., Park, J. H., Lee, E. J., Vo, T. T., Choi, H., Kim, J. Y., et al. (2016). ARD1-mediated Hsp70 acetylation balances stress-induced protein refolding and degradation. *Nature Communications, 7*, 12882.

Sers, C., Kuner, R., Falk, C. S., Lund, P., Sueltmann, H., Braun, M., et al. (2009). Downregulation of HLA Class I and NKG2D ligands through a concerted action of MAPK and DNA methyltransferases in colorectal cancer cells. *International Journal of Cancer, 125*, 1626–1639.

Shen, L., Orillion, A., & Pili, R. (2016). Histone deacetylase inhibitors as immunomodulators in cancer therapeutics. *Epigenomics, 8*, 415–428.

Sumimoto, H., Takano, A., Teramoto, K., & Daigo, Y. (2016). RAS-mitogen-activated protein kinase signal is required for enhanced PD-L1 expression in human lung cancers. *PLoS One, 11*, e0166626.

Terranova-Barberio, M., Thomas, S., & Munster, P. N. (2016). Epigenetic modifiers in immunotherapy: A focus on checkpoint inhibitors. *Immunotherapy, 8*, 705–719.

Tiper, I. V., & Webb, T. J. (2016). Histone deacetylase inhibitors enhance CD1d-dependent NKT cell responses to lymphoma. *Cancer Immunology, Immunotherapy, 65*, 1411–1421.

Vo, D. D., Prins, R. M., Begley, J. L., Donahue, T. R., Morris, L. F., Bruhn, K. W., et al. (2009). Enhanced antitumor activity induced by adoptive T-cell transfer and adjunctive use of the histone deacetylase inhibitor LAQ824. *Cancer Research, 69*, 8693–8699.

Watamoto, K., Towatari, M., Ozawa, Y., Miyata, Y., Okamoto, M., Abe, A., et al. (2003). Altered interaction of HDAC5 with GATA-1 during MEL cell differentiation. *Oncogene, 22*, 9176–9184.

West, A. C., & Johnstone, R. W. (2014). New and emerging HDAC inhibitors for cancer treatment. *The Journal of Clinical Investigation, 124*, 30–39.

West, A. C., Mattarollo, S. R., Shortt, J., Cluse, L. A., Christiansen, A. J., Smyth, M. J., et al. (2013). An intact immune system is required for the anticancer activities of histone deacetylase inhibitors. *Cancer Research, 73*, 7265–7276.

Yang, H., Chen, H., Luo, S., Li, L., Zhou, S., Shen, R., et al. (2017). The correlation between programmed death-ligand 1 expression and driver gene mutations in NSCLC. *Oncotarget, 8*, 23517–23528.

Yang, H., Lan, P., Hou, Z., Guan, Y., Zhang, J., Xu, W., et al. (2015). Histone deacetylase inhibitor SAHA epigenetically regulates miR-17-92 cluster and MCM7 to upregulate MICA expression in hepatoma. *British Journal of Cancer, 112*, 112–121.

Yang, P. H., Zhang, L., Zhang, Y. J., Zhang, J., & Xu, W. F. (2013). HDAC6: Physiological function and its selective inhibitors for cancer treatment. *Drug Discoveries & Therapeutics, 7*, 233–242.

Yuan, Z. L., Guan, Y. J., Chatterjee, D., & Chin, Y. E. (2005). Stat3 dimerization regulated by reversible acetylation of a single lysine residue. *Science, 307*, 269–273.

Zhan, P., Wang, X., Liu, X., & Suzuki, T. (2017). Medicinal chemistry insights into novel HDAC inhibitors: An updated patent review (2012–2016). *Recent Patents on Anti-Cancer Drug Discovery, 12*, 16–34.

Zhang, W., Pang, Q., Yan, C., Wang, Q., Yang, J., Yu, S., et al. (2017). Induction of PD-L1 expression by epidermal growth factor receptor-mediated signaling in esophageal squamous cell carcinoma. *OncoTargets and Therapy, 10*, 763–771.

Zheng, H., Zhao, W., Yan, C., Watson, C. C., Massengill, M., Xie, M., et al. (2016). Inhibitors enhance T-cell chemokine expression and augment response to PD-1 immunotherapy in lung adenocarcinoma. *Clinical Cancer Research, 22*, 4119–4132.

Zhou, Q., Agoston, A. T., Atadja, P., Nelson, W. G., & Davidson, N. E. (2008). Inhibition of histone deacetylases promotes ubiquitin-dependent proteasomal degradation of DNA methyltransferase 1 in human breast cancer cells. *Molecular Cancer Research*, *6*, 873–883.

## FURTHER READING

Mazzone, R., Zwergel, C., Mai, A., & Valente, S. (2017). Epi-drugs in combination with immunotherapy: A new avenue to improve anticancer efficacy. *Clinical Epigenetics 9*, 59. https://doi.org/10.1186/s13148-017-0358-y.

Schiffmann, I., Greve, G., Jung, M., & Lübbert, M. (2016). Epigenetic therapy approaches in non-small cell lung cancer: Update and perspectives. *Epigenetics*, *11*, 858–870.

CHAPTER TWO

# A Theoretical Basis for the Efficacy of Cancer Immunotherapy and Immunogenic Tumor Dormancy: The Adaptation Model of Immunity

Masoud H. Manjili[*,†,1]
[*]VCU Institute of Molecular Medicine, Virginia Commonwealth University, School of Medicine, Richmond, VA, United States
[†]VCU Massey Cancer Center, Virginia Commonwealth University, School of Medicine, Richmond, VA, United States
[1]Corresponding author: e-mail address: masoud.manjili@vcuhealth.org

## Contents

1. Introduction    18
2. Outcome of Cancer Immunotherapies Inspired by the SNS and Danger Models    19
    2.1 Targeting Tumor-Associated Antigens or Tumor-Specific Antigens?    19
    2.2 Allogeneic Cancer Vaccines    20
    2.3 Allogeneic Stem Cell Transplantation    21
    2.4 Neoantigen Cancer Vaccines and Engineered TcR    22
3. Beyond the SNS and Danger Models: Tumor Escape and Immune Evasion    22
4. Discovery and Modulation of Tumor Adaptation Receptors    24
    4.1 Central Tolerance and the Adaptation Model    24
    4.2 ARs and ALs: (i) The Endothelin Axis    25
    4.3 ARs and ALs: (ii) The PD-L1/PD-1 Checkpoint Pathway    27
5. Immunogenic Dormancy of Occult Tumor Cells Through Adaptation    29
Acknowledgments    31
Financial Support    31
Disclosures    31
References    32

## Abstract

In the past decades, a variety of strategies have been explored to cure cancer by means of immunotherapy, which is less toxic compared with chemotherapy or radiation therapy, and could establish memory for long-lasting protection against tumor recurrence. These endeavors have been successful in offering therapeutic antibodies, vaccines, or cellular immunotherapies, which resulted in prolonging survival of some cancer patients; however, complete cures have not been consistently achieved. The conception, design, and implementation of these promising immunotherapeutic strategies have been influenced by two schools of thought in immunology, which include the "self–nonself" (SNS) model and the "danger" model. Further progress in cancer immunotherapy to achieve consistent cancer cures requires an evolution in our understanding of how the immune system works. The purpose of this review is to revisit premises and limitations of the SNS and danger models based on the outcomes of cancer immunotherapies by suggesting that both models are two sides of the same coin describing how the immune response is induced against cancer. However, neither explains how the immune response succeeds or fails in eliminating the tumor. To this end, the adaptation model has been proposed to explain efficacy of the immune response for achieving cancer cure.

## 1. INTRODUCTION

The "self–nonself" (SNS) model (Janeway, 1992) and the "danger" model (Matzinger, 2002) of immunity appear to be on opposite sides of thought in describing how the immune system functions. However, growing evidence suggests that both concepts are complementary when it comes to describing how an immune response is induced against cancer rather than how it succeeds or fails to eliminate cancer. For an antitumor immune response, T cells must receive two signals. Signal I is provided by the presentation of tumor antigens to T cells in the context of major histocompatibility complex/T cell receptor (MHC/TcR) interaction, and signal II is provided by T helper cells (Bretscher & Cohn, 1970) or costimulatory molecules such as B7.1/B7.2-CD28 (Janeway, 1992). Although the original SNS model (Bretscher & Cohn, 1970) does not have an explanation for signal II, an evolved version of the SNS model suggests that signal II is also induced by foreign proteins recognized by pathogen-associated molecular patterns (PAMPs) on the immune cells (Janeway, 1992). However, PAMPs such as toll-like receptors (TLR) also recognize self-proteins or endogenous ligands (Yu, Wang, & Chen, 2010). In some classifications, cytokine signaling during T cell activation or differentiation is considered as signal III; however, the proposed classification is that signals I and II are involved in T cell activation and differentiation. Therefore, both costimulatory molecules and cytokine signaling are considered as signal II. The SNS model solely emphasizes foreignness and focuses on the affinity of T cell receptor for the antigen. This model proposes that foreign antigens usually have a stronger affinity for T cell activation because self-antigen-educated T cells develop tolerance in the thymus. The danger model emphasizes on danger signals in response to any damage being harmful to the host and which induces signal II. The

# The Adaptation Model of Immunity

**Table 1** Three Signals During Antitumor Immune Responses

| Models | Signals | Molecules | Function | Outcomes |
|---|---|---|---|---|
| SNS | Signal I | MHC-TcR | Antigen recognition | T cell activation and differentiation |
| Danger | Signal II | B7.1/B7.2-CD28 | T cell activation | |
| | | Cytokines | T cell differentiation | |
| Adaptation | Signal III | AR-AL | T cell function | Success or failure of the immune response |

danger signals include damage-associated molecular pattern; PAMP could also be considered as danger signal because of being expressed on pathogens that are harmful to humans. Without signal II, signal I induces tolerance toward antigens. In fact, the danger model is the evolution of the SNS model by theorizing the entity of signal II in the induction of the immune response regardless of the self or nonself entity of signal I, the antigen. The evolutionary relationship between the SNS model and the danger model is similar to that of tumor immunosurveillance and tumor immunoediting theories (Dunn, Bruce, Ikeda, Old, & Schreiber, 2002). Vaccines have been designed based on the inspiration from the SNS model by including highly immunogenic antigens as signal I, and from the danger model by including adjuvants, regardless of the self or nonself entity of adjuvants, to induce signal II. To understand how an antitumor immune response succeeds or fails in eliminating the tumor, a signal III has to be involved. Signal III is a communication signaling that determines whether tumor cells die, proliferate, or become dormant following vaccination or immunotherapy (Table 1). The adaptation model proposes that this communication signaling has to be orchestrated through adaptation receptors (ARs) and adaptation ligands (ALs) that are distinct from costimulation (Manjili, 2014).

## 2. OUTCOME OF CANCER IMMUNOTHERAPIES INSPIRED BY THE SNS AND DANGER MODELS

### 2.1 Targeting Tumor-Associated Antigens or Tumor-Specific Antigens?

The SNS model suggests that the sequence or nature of tumor antigens determines the strength of an antitumor immune response. Whereas tumor-associated antigens (TAAs) are thought to be weakly immunogenic, tumor-specific antigens (TSAs) are considered to be highly immunogenic. This assumption is based on the SNS model without empirical evidence demonstrating that immunotherapeutic targeting of TSAs or foreign-like antigens is more effective than that of targeting TAAs or self-antigens. Although targeting mutant neoantigens is a viable immunotherapeutic strategy supported by the SNS model, it is not more effective than targeting TAAs. To target TAAs or TSAs in a vaccine formulation, the danger model provides a conceptual framework emphasizing the use of an adjuvant in order to induce signal II (Gallucci, Lolkema, & Matzinger, 1999). The danger model suggests that the use of an effective adjuvant and continuous vaccination is important for antitumor efficacy of a vaccine (Gallucci et al., 1999; Matzinger, 2002). Immunotherapeutics that target TAAs have been approved by the FDA based on prolonging survival of patients with carcinomas when used in a therapeutic setting. For instance, prostatic acid phosphatase is a TAA being used in sipuleucel-T (Provenge) vaccine against asymptomatic or minimally symptomatic metastatic hormone refractory prostate cancer, and extended survival of patients by a median of 4.1 months (Kantoff et al., 2010). HER2/neu is another TAA being used as a target for antibody therapy of metastatic breast cancer. Addition of anti-HER2/neu antibody therapy to chemotherapy prolonged a median survival of 5.1 months (Slamon et al., 2001). Two FDA-approved HPV and EBV vaccines containing TSAs—nonself viral antigens—have been tested in prophylactic settings for the prevention of cervical cancer and liver cancer, respectively. Importantly, the efficacy of these vaccines has more to do with their use in prophylactic settings, rather than the nature of the antigen being foreign entity or an adjuvant being a strong inducer of danger signals.

## 2.2 Allogeneic Cancer Vaccines

To enhance immunogenicity of cancer vaccines, an allogeneic system has been designed and tested in a randomized phase III clinical trial using Canvaxin (Kelland, 2006). The vaccine consists of allogeneic, living whole melanoma cells, as a source of foreign antigens, and BCG as adjuvant. According to the SNS model, the inclusion of foreign antigens (Bretscher & Cohn, 1970) and a foreign adjuvant (Janeway, 1992) was expected to induce robust antitumor immune responses. However, the trial was discontinued prematurely because survival benefit was unlikely to be achieved (Kelland, 2006). Another allogeneic vaccine called GVAX (Cell Genesys, Inc.) consisting of allogeneic pancreatic cancer cell lines transfected with a human *GM-CSF* gene as adjuvant. GVAX was tested in combination with CTLA4 blockade in patients with previously treated advanced pancreatic ductal adenocarcinoma and resulted in prolonging a median overall survival of only 5.7 vs 3.6 months for CTLA4 schedule alone (Le et al., 2013). However, no complete cures were achieved. It has been suggested that the inclusion of foreign helper epitopes should be sufficient to induce an effective antitumor CD8+ T cell response (Anderson, 2014) without overloading the immune system with foreign antigens. Despite an improved efficacy, this strategy did not provide a complete protection against the tumor in animal models (Snook, Magee, Schulz, & Waldman, 2014; Steinaa, Rasmussen, Rygaard, Mouritsen, & Gautam, 2007).

## 2.3 Allogeneic Stem Cell Transplantation

Allogeneic stem cell transplantation (SCT) is a promising immunotherapeutic approach for the treatment of patients with hematological malignancies. This strategy is based on the SNS model, proposing that donor T cells will recognize recipient tumor cells as nonself entities and attack them. The treatment has to be performed in the setting of donor recipient being matched in major histocompatibility antigens, HLA-A, -B, -C, DR, and ideally DQ. However, mismatch in minor histocompatibility antigens could induce an alloreactive immune response, which is often associated with graft vs host disease (GVHD). Allogeneic SCT is usually given along with irradiation or chemotherapy to the recipient, which could potentially function as adjuvant depending on the immunogenic nature of some chemotherapies or radiation therapies at certain doses. The danger model proposes that signal II is readily induced in organs such as the skin and the gut because these organs are exposed to the external world, commensals and pathogens, which cause damage and induce danger (Matzinger, 2012); this could act as adjuvant or danger for allogeneic SCT and result in GVHD in these organs. However, these alarming conditions also exist in recipients of autologous SCT without causing severe GVHD. What has been less appreciated is the role of conditioning regimens in disrupting homeostatic cellular adaptation that contributes to the development of tissue-specific GVHD (Manjili & Toor, 2014). Treatment for GVHD is also inspired by the SNS model, assuming that alloreactive T cells are responsible for GVHD; therefore, immunosuppressive drugs are given as GVHD prophylaxis or as therapeutic regimens, rendering patients susceptible to infections and increasing the risk of tumor relapse. The SNS model has not been able to offer an effective therapeutic strategy for GVHD without compromising the patient immune response. The danger model suggests that the high frequency of GVHD in the gut, the skin, and the liver is because these organs are most in contact with commensals and pathogens producing danger signals. That is why allogeneic SCT fails to induce severe GVHD in germ-free animals (Matzinger, 2012). However, similar danger signals are present in the gut, the skin, and the liver following autologous SCT without causing a severe GVHD.

## 2.4 Neoantigen Cancer Vaccines and Engineered TcR

The next-generation cancer vaccines that have been conceived based on the SNS and danger models contain mutant neoantigens and adjuvant. The idea is based on the understanding that cancer cells usually undergo somatic mutations resulting in the expression of mutant antigens that can be considered as nonself, because they are not expressed during central tolerance. Mutant tumor antigens have been detected in cancer patients (Assadipour et al., 2017; Verdegaal et al., 2016), though they do not induce tumor rejection. Vaccination with defined neoantigens in combination with poly I:C adjuvant has shown some efficacy in mice when combined with immune-checkpoint inhibitors (Gubin et al., 2014). Thus far, no human data are available to confirm antitumor efficacy of neoantigen vaccines. Another immunotherapeutic strategy inspired by the SNS model is enhancing affinity of T cells for target antigens by means of engineered TcR. This strategy can be combined with targeting neoantigens. A combination of two strategies by targeting KRAS-mutant neoantigens and using T cells engineered to express TcR specific for the appropriate KRAS mutations was elegantly tested in mice (Wang et al., 2016). Adoptive transfer of the KRAS-mutant-specific transduced T cells significantly reduced pancreatic tumor growth in nonobese diabetic scid gamma mice, but the treatment did not eliminate the tumors (Wang et al., 2016). Such outcomes have been attributed to the neoantigen immunoediting by T cells, and it was suggested that induction of broad neoantigen-specific T cell responses should be used to avoid tumor resistance (Verdegaal et al., 2016).

## 3. BEYOND THE SNS AND DANGER MODELS: TUMOR ESCAPE AND IMMUNE EVASION

Immunotherapeutic strategies that have been inspired by the SNS and danger models have shown limited efficacy against cancer. Such outcomes have been attributed to tumor escape and immune evasion, which cannot be directly explained by either the SNS or danger models. In fact, these models can explain the induction of the immune response rather than predicting its outcome. To overcome a single tumor antigen loss, multiple tumor antigens have been used and epigenetic modulators have been tested to induce the expression of a panel of cancer testis antigens (CTAs) so as to overcome a single antigen loss during immunotherapy. A randomized phase II clinical trial of multiepitope vaccine in patients with stage IV melanoma increased median overall survival by a few months (Slingluff et al., 2013). A combination of decitabine to induce CTAs and a vaccine targeting NY-ESO1 in ovarian cancer resulted in a partial response (Odunsi et al., 2014). In patients with stage IV melanoma, a combination vaccine comprised of six HLA-DR-restricted peptides increased median overall survival of 4.1 years compared with control arm (Hu, Kim, Blackwell, & Slingluff, 2015). Immune evasion mechanisms have also been targeted by various strategies. For instance, tumor-induced immunosuppressive cells such as regulatory T cells (Tregs) and myeloid-derived suppressor cells (MDSCs) have been targeted in combination with immunotherapy, yet cancer cure has not been achieved. In patients with head and neck squamous cell carcinoma, tadalafil treatment significantly reduced both MDSCs and Tregs and increased tumor-specific immune responses, but no objective response was reported (Weed et al., 2015). In the 4T1 murine mammary tumor model, decitabine combined with adoptive immunotherapy (AIT) resulted in tumor inhibition and an increased rate of cure (Terracina et al., 2016), though its therapeutic efficacy against locally advanced tumor or established tumor metastasis has not been shown. In an animal model of HER2/neu-positive mammary carcinoma, depletion of MDSCs and induction of the expression of a panel of CTAs by decitabine, combined with AIT, resulted in prolonging survival of animals carrying metastatic breast cancer in the lung, although animals eventually succumbed to the tumor (Payne et al., 2016). In addition, targeting immune-checkpoint pathways of immune evasion by using anti-CTLA4 or anti-PD-1/PD-L1 antibody resulted in prolonging survival of cancer patients (Achkar & Tarhini, 2017), but again, a consistent and complete remission has yet to be achieved. Therefore, tackling several tumor escape pathways during immunotherapeutic regimens that were inspired by the SNS model or the danger model could improve the clinical outcome for cancer patients but could not consistently achieve a cancer cure. A continuous immunization, as suggested by Matzinger, may maintain antitumor immune responses, but it could not offer a cure for cancer because of tumor escape mechanisms.

## 4. DISCOVERY AND MODULATION OF TUMOR ADAPTATION RECEPTORS

The adaptation model of immunity was recently proposed to explain efficacy of the immune response during cancer, infectious diseases, allergy, and autoimmune diseases (Manjili, 2014). The model proposes a different theoretical perspective in tumor immunology and immunotherapy by suggesting that dysregulation of target tissues for the expression of ARs and ALs renders them susceptible or resistant to ongoing immune responses.

### 4.1 Central Tolerance and the Adaptation Model

Positive selection results in the maturation of CD4+CD8+ T cells into a single-positive CD4+ or CD8+ T cells via MHC class II or MHC class I restriction, respectively. During positive selection, MHC/self-peptide complex (signal I) selects and supports survival of T cells that are self-reactive. However, the affinity of these T cells for self-antigens is low due to the nature of cortical thymic epithelial cells (cTECs) expressing wobbly or private peptides that bind MHC molecules weakly. The cTECs express β5t-containing thymoproteasomes, which inefficiently cleave substrates adjacent to hydrophobic amino acids of self-peptides, and as a result create wobbly binding of β5t-derived peptides with a faster TcR off-rate (Murata et al., 2007; Ziegler, Muller, Bockmann, & Uchanska-Ziegler, 2009). On the other hand, medullary TEC or DCs express β5i-containing immunoproteasomes, which are efficient in cleaving substrates adjacent to hydrophobic amino acids and create high-affinity MHC/self-peptides for all positively selected T cells. Therefore, similar peptides can have different affinities during positive and negative selections. Medullary DCs also express costimulatory molecules such as CD40, B7-1, and B7-2 (signal II) (Klein, Hinterberger, Wirnsberger, & Kyewski, 2009). Around two-thirds of medullary DCs are CD11c$^{high}$ DCs, which contain CD8α+ thymic resident DCs, which are efficient in antigen cross-presentation, and CD8α− migratory DCs (Li, Park, Foss, & Goldschneider, 2009). Medullary DCs express a wide array of tissue-specific antigens regulated by the autoimmune regulator (*AIRE*) gene as well as *AIRE*-independent mechanisms (Derbinski, Schulte, Kyewski, & Klein, 2001; Takaba et al., 2015). Negative selection is a mystery that has not been fully understood by the SNS or the danger model. A classical explanation is that T cells die because of the high affinity for antigens, while those with a low-affinity survive. This explanation raises some questions: (i) theoretically, all positively selected T cells recognizing β5t-derived peptides should have a higher affinity for the β5i-derived peptides in the medulla, so why do some T cells die and some survive during negative selection? (ii) Why do high-affinity T cells die upon activation in the thymus, but they survive in the periphery? T cells that were matured from double-positive into single-positive cells in the cortex should function like alloreactive T cells after activation upon recognizing high-affinity antigens. T cell activation also takes place in the medulla in the absence of any danger signals; (iii) why do surviving T cells not get activated upon receiving signal I and signal II in the medulla, but they do get activated in the periphery? The β5i-containing immunoproteasomes in the medulla increase the affinity of self-peptides for surviving T cells, while they also receive signal II, yet they do not get activated. The adaptation model (Manjili, 2014) proposes that negatively selected T cells in the medulla express ARs and thus survive upon antigen recognition, whereas defective T cells that lack ARs will be eliminated upon antigen recognition; if these T cells escape from negative selection, they would die in the periphery upon activation. Therefore, the purpose of negative selection is to eliminate faulty T cells and select functional T cells that are able to survive upon activation. Autoreactive T cells could not be the otherwise deleted T cells because thymic emigration decreases in AIRE$^{-/-}$ mice (Jin et al., 2017), suggesting that autoreactivity is not because of the escape of otherwise deleted T cells and their addition to the pool of surviving T cells. On the other hand, autoreactive T cells are perhaps those that do not die during negative selection in spite of recognizing MHC/self-antigens. In the periphery, upon engagement of ALs on DCs with ARs on T cells during activation, ARs transduce survival signals in T cells by inducing the expression of antiapoptotic proteins, such as cFLIP and Bcl-xL (Paulsen & Janssen, 2011). Lack of expression of ALs by APCs could also result in activation-induced cell death (AICD) in T cells. For instance, hepatic DCs induce apoptosis in T cells during activation, whereas splenic DCs support survival of activated T cells (Bertolino, Trescol-Biemont, & Rabourdin-Combe, 1998).

## 4.2 ARs and ALs: (i) The Endothelin Axis

Cancer patients often harbor preexisting antitumor immune responses that fail to protect the patients from cancer (Lu et al., 2012). Also, immunotherapy as a single agent often fails to eliminate the tumor. Similar observations were made in different diseases. For instance, healthy individuals and patients with multiple sclerosis (MS) harbor T cells that recognize myelin basic proteins (MBPs), but a pathogenic manifestation of the immune response is evident only in MS patient (Martin, Whitaker, Rhame, Goodin, & McFarland, 1994). Similarly, preexisting anti-DNA autoantibodies were detected in healthy individuals and patients with lupus erythematosus with a pathogenic manifestation only in the latter (Martin et al., 1994). Th1 and Th17 inflammatory cells in the gut can protect the host from *Helicobacter pylori* infection without any toxicity to the tissue (Ding et al., 2013), but they become destructive during Crohn's disease. These paradoxical observations suggest that the immune response alone is not the primary factor in the pathogenesis of autoimmune diseases or inefficacy in cancer patients; rather, alterations in the expression of AR on the target cells could render them susceptible or resistant to the immune response. In fact, an altered gut microbiome profile is associated with Crohn's disease such that nutritional therapy can modulate pediatric Crohn's disease (de la Cruz-Merino et al., 2011), again suggesting that gut microbiome is an important factor in regulating the expression of ARs in the tissue. Tumor cells that arise from normal cells, perhaps, retain their ARs to survive immune surveillance. One candidate for the AR/AL is the endothelin axis, which includes the endothelin (ET) containing ET-1, ET-2, and ET-3 isoforms as ALs, and the ET receptor A ($ET_A$) as an AR. Activation of the $ET_A$ AR by the ET-1 AL can lead to the induction of survival pathways, whereas activation of the $ET_B$, which antagonizes the $ET_A$, results in apoptosis (Nelson, Udan, Guruli, & Pflug, 2005). ETs are expressed by a variety of cell types including endothelial cells, macrophages, astrocytes, and neurons (Simonson, 1993). The $ET_A$ receptor has a greater affinity for ET-1, and the $ET_B$ receptor binds to all three ET isoforms equally (Arai, Hori, Aramori, Ohkubo, & Nakanishi, 1990). ET-1 is upregulated by astrocytes in a number of brain pathologies, including MS (D'haeseleer et al., 2013) and Alzheimer's disease (Palmer, Barker, Kehoe, & Love, 2012), as well as in rheumatoid arthritis (Haq, El-Ramahi, Al-Dalaan, & Al-Sedairy, 1999) and cancer (Wulfing et al., 2004). $ET_B$ is upregulated in active MS lesions (Yuen et al., 2013), and ET-1 acts almost exclusively through $ET_B$, and not $ET_A$, on astrocytes to inhibit remyelination (Hammond et al., 2015). Therefore, it is reasonable to predict that alterations in the balance between the $ET_A$ AR and its antagonist receptor, the $ET_B$, render the nervous system susceptible to anti-MBP immune responses. In humans, $ET_A$ acts as an AR by inducing the expression of antiapoptotic genes in prostate cancer (Nelson et al., 2005). Its ligand, ET-1, acts as an AL and is produced by the prostate epithelia (Nelson et al., 2005). The ET-1/$ET_A$ pathway is involved in the inhibition of apoptosis in melanocytes during UV irradiation (Swope & Abdel-Malek, 2016). In fact, a higher responsiveness of melanoma patients to immunotherapy compared with patients with prostate cancer or ovarian cancer could be because the $ET_A$ AR is upregulated in prostate and ovarian cancers but not in melanoma (Nelson, Bagnato, Battistini, & Nisen, 2003). The ET-1 AL is produced by the prostate epithelia (Nelson et al., 2005); in prostate cancer, not only a key component of ET-1 clearance, the $ET_B$ receptor, is diminished (Nelson et al., 1996), but also the $ET_A$ AR is upregulated (Nelson et al., 2003). These could make tumor-infiltrating T cells ineffective in patients with prostate cancer. Human DCs also produce ET-1 upon activation (Spirig et al., 2009), which in turn support survival of T cells during activation as well as tumor cells that express $ET_A$. ET-1 is also involved in the survival of activated T cells during autoimmune systemic sclerosis (Elisa et al., 2015). In rats, the ET-1/$ET_A$ pathway is critical for thymocyte proliferation (Malendowicz, Brelinska, De Caro, Trejer, & Nussdorfer, 1998).

## 4.3 ARs and ALs: (ii) The PD-L1/PD-1 Checkpoint Pathway

The programmed cell death-1 (PD-1) receptor is expressed on activated T cells. Its ligands, PD-L1 and PD-L2, are commonly expressed on dendritic cells or macrophages. PD-L1 is a bidirectional membrane protein acting as a ligand to induce anergy in PD-1-positive T cells and acting as an AR to induce antiapoptotic genes in PD-L1-positive target cells (Azuma et al., 2008). Constitutive expression of PD-L1 in the immune-privileged sites such as cornea and retina protects them from GVHD following corneal allograft, despite infiltration of CD4+ T cells; however, blockade of PD-L1 accelerates allograft rejection (Hori et al., 2006). In a murine model, PD-L1 deficiency in pancreatic beta-cells triggers their destruction by CD8+ T cells (Rajasalu et al., 2010). An altered expression of PD-L1 correlates with not only autoimmune diseases but also cancer progression. For instance, PD-L1 loss was reported in children with systemic lupus erythematosus, and expression of PD-L1 is restored only during disease remission (Mozaffarian, Wiedeman, & Stevens, 2008). The expression of PD-L1 on activated T cells supports their survival such that PD-L1-deficient T cells express lower Bcl-xL, which is an antiapoptosis gene, than wild-type cells and are more sensitive to apoptosis in vivo (Pulko et al., 2011). Tumor cells exploit this pathway by the expression of PD-L1 in order to survive immune surveillance. Antitumor T cells can upregulate PD-L1 on tumor cells through the production of IFN-γ. For instance, upregulation of PD-L1 is only detected in tumor cells that are adjacent to IFN-γ-producing TILs in melanoma patients (Taube et al., 2012). Of note, tumor cells also utilize IFN-γ-independent pathways for the expression of PD-L1 which involve *PTEN* (Parsa et al., 2007) or EGFR (Akbay et al., 2013). In phase I clinical trial, anti-PD-1 therapy showed cumulative response rates of 18%, 28%, and 27% among patients with non-small-cell lung cancer, melanoma, and renal cell carcinoma, respectively (Topalian et al., 2012). More recently, an objective response rate of 30%–40% in melanoma patients has been reported (Robert et al., 2015; Topalian et al., 2014). A high variety of response rates among different types of cancers to PD-1 immune-checkpoint inhibition therapy suggest the involvement of additional ARs that support tumor cell survival when the PD-L1 pathway is blocked. According to the adaptation model of immunity, antitumor efficacy of anti-PD-1/PD-L1 immunotherapy as a single agent is mainly due to the blockade of antiapoptotic gene expression downstream of PD-L1 on tumor cells. Therefore, the model predicts a higher efficacy of anti-PD-L1 therapy than anti-PD-1 therapy. In fact, some types of anti-PD-L1 antibodies can inhibit the interaction of not only PD-L1 and PD-1 but also PD-L1 and CD80 (Keir, Butte, Freeman, & Sharpe, 2008). On the other hand, blockade of PD-1 can rescue effector T cells from suppression, but the engagement of PD-L1 on tumor cells with CD80 on APCs can still induce survival signaling in tumor cells, facilitating resistance of tumor cells to antitumor effector T cells (Fig. 1). However, studies performed

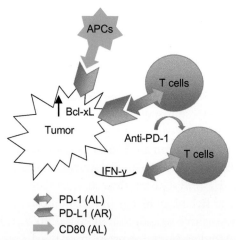

**Fig. 1** PD-L1 acts as an AR on tumor cells. Anti-PD-1 could block PD-1/PD-L1 interaction and result in rescuing T cells from suppression. However, the engagement of CD80 on APCs with PD-L1 on tumor cells can upregulate the antiapoptotic gene Bcl-xL in tumor cells and support their survival in the presence of IFN-γ-producing antitumor T cells.

in the context of SNS model pay more attention to rescuing T cells from the suppression rather than blocking survival signaling in tumor cells following anti-PD-1/PD-L1 immunotherapy.

## 5. IMMUNOGENIC DORMANCY OF OCCULT TUMOR CELLS THROUGH ADAPTATION

An effective antitumor immune response, which is capable of inducing tumor regression, cannot guarantee elimination of tumor dormancy. In fact, immune responses induce the expression of an AR, PD-L1, on tumor cells through secretion of IFN-γ (Payne et al., 2016). IFN-γ is a dual-edged cytokine capable of inducing apoptosis and also facilitating tumor dormancy (Liu et al., 2017). Immunogenic tumor dormancy has been documented during unintentional transplantation of cancer into immunocompromised recipients from organ donors who were in clinical remission (Kauffman, McBride, & Delmonico, 2000) or with no clinical history of cancer (Myron Kauffman et al., 2002). Immunogenic tumor dormancy is defined by the expression of mutant antigens, increased MHC-I, cell membrane translocation of calreticulin, release of ATP, release of nonhistone chromatin-binding protein high-mobility group box 1, and secretion of immunostimulatory cytokines such as type I interferons (Michaud et al., 2011, 2014; Sistigu et al., 2014). A mechanism of immunogenic tumor dormancy was demonstrated in an animal model of methylcholanthrene-induced sarcoma (Koebel et al., 2007). Immunogenic dormancy is also evident in *Mycobacterium tuberculosis* infection keeping the infectious agent in dormant or latent state, thus protecting the host from active disease. Long latency before the appearance of AIDS is also evident in the presence of the immune response (Goonetilleke et al., 2009). HIV-infected CD4+ T cells express PD-L1 (Trabattoni et al., 2003), which could be kept dormant by HIV-specific PD-1$^{low}$ CD8+ T cells during the latency period. Whereas PD-1$^{high}$ effector T cells can be suppressed through PD-L1 engagement allowing tumor growth, the PD-1$^{low}$ effector T cells could remain active and push PD-L1-positive tumor cells into the state of immunogenic dormancy by producing IFN-γ; dormant tumor cells will remain in check by the immune response until they escape from dormancy. Thus far, two types of tumor dormancy have been reported, which include Ki67$^-$ quiescent dormancy and Ki67$^{low}$ indolent dormancy (Payne et al., 2016). Similar to actively proliferating tumor cells, the indolent, but not quiescent, dormant cells can evolve through immunoediting and escape from the immune response. Recently, an elegant study by Dr. Restifo's group demonstrated that tumor necrosis releases an intracellular ion, potassium, into the extracellular fluid at the tumor site and results in the suppression of effector T cells. They showed that ionic reprogramming of tumor-specific T cells can improve their effector functions and prolong survival of melanoma-bearing mice (Eil et al., 2016). In clinical settings, targeting neoantigens by immunotherapy resulted in the stabilization of metastatic cholangiocarcinoma for 13 months, and then, disease progression was observed in the lungs (Tran et al., 2014). In a separate study, adoptive T cell therapy using a polyclonal CD8+ TIL recognizing mutant KRAS G12D in a patient with metastatic colorectal cancer resulted in the regression of lung metastatic lesions. However, one lesion escaped through loss of heterozygosity of the copy of chromosome 6 that encoded HLA-C*08:02 (Tran et al., 2015; Tran et al., 2016). Complete regression of neu-overexpressing mammary carcinoma and subsequent relapse of antigen-negative tumor variant have been reported in a semiallogeneic model in which T cells and tumor cells were matched in major but not minor histocompatibility antigens (Kmieciak, Knutson, Dumur, & Manjili, 2007; Santisteban et al., 2009). Effectiveness of immunotherapy in some cancer patients but not others perhaps results from differences in the expression of ARs and/or ALs regulated by different oncogenes or epigenetic alterations. The adaptation model can also explain sterile chronic inflammation where the immune response to self-antigens is induced in the presence of signals I and II, but rather than destroying target organs, it initially inhibits cell growth because of the presence of ARs on target tissues, and eventually facilitates escape of natural malignant cells from dormancy (Manjili, 2017). Advances in our understanding of the AR/AL pathways are expected to lead to a breakthrough in immunotherapeutic treatment of cancer.

In summary, the adaptation model of immunity proposes that the status of ARs/ALs on tumor cells and T cells, respectively, determines the outcome of antitumor immune responses. There are four scenarios predicted by the adaptation model of immunity (Table 2). Tumor cells expressing ARs (ARs+) will receive survival signals from T cells by engaging with ALs on T cells (ALs+) and as a result become dormant as long as antitumor effector T cells are present. Other tumor-infiltrating cells such as myeloid cells could also express PD-1. Also, tumor cells expressing ALs will induce survival signals in effector T cells that express ARs (Scenario 1). Alterations in the expression of ARs/ALs on tumor cells could change the outcome, leading to the elimination of tumor cells that lack ARs (ARs−) by effector T cells (Scenarios 2). Tumor cells that do not express ALs fail to induce

Table 2 Outcomes of Antitumor Immune Responses

| Scenarios | Tumor | Effector T Cells | Outcomes |
|---|---|---|---|
| 1 | AR+ | AL+ | Tumor dormancy |
|   | AL+ | AR+ | T cell survival |
| 2 | AR− | AL+/− | Tumor elimination |
|   | AL+ | AR+ | T cell survival |
| 3 | AR− | AL+/− | Tumor elimination |
|   | AL− | AR+/− | T cells undergo AICD |
| 4 | AR+ | AL+ | Tumor escape and relapse |
|   | AL− | AR−/+ | T cells undergo AICD |

survival signals in antitumor T cells, and these T cells will undergo AICD if they do not receive survival signals from stromal cells (Scenario 3). Finally, dormant tumor cells could escape from the immune response by down-regulating the expression of ALs on antitumor T cells (ALs −) and relapse (Scenario 4). Advances in our understanding and identification of ARs and ALs could lead to targeted therapies for epigenetic silencing of ARs on tumor cells, thereby rendering them vulnerable to immunotherapy.

## ACKNOWLEDGMENTS

My sincere apology to those whose invaluable contributions to the field were not cited due to space limitations. I gratefully acknowledge the support of VCU Massey Cancer Center and the Commonwealth Foundation for Cancer Research. My special thanks go to Elizabeth Repasky and Scott Abrams of Roswell Park Cancer Institute, Anthony Vella of UCONN Health, Nejat Egilmez of the University of Louisville, Colin Anderson of the University of Alberta, and Harry Bear of Virginia Commonwealth University Massey Cancer Center, whose feedback helped clarify my argument.

## FINANCIAL SUPPORT

This work was supported by the Office of the Assistant Secretary of Defense for Health Affairs through the Breast Cancer Research Program under Award No. W81XWH-14-1-0087, and a pilot funding from the VCU Massey Cancer Center supported, in part, with funding from NIH/NCI Cancer Center Support Grant P30 CA016059.

## DISCLOSURES

Opinions, interpretations, conclusions, and recommendations are those of the authors and are not necessarily endorsed by the US Department of Defense. The authors declare no conflicts of interest.

# REFERENCES

Achkar, T., & Tarhini, A. A. (2017). The use of immunotherapy in the treatment of melanoma. *Journal of Hematology & Oncology, 10*(1), 88. https://doi.org/10.1186/s13045-017-0458-3.

Akbay, E. A., Koyama, S., Carretero, J., Altabef, A., Tchaicha, J. H., Christensen, C. L., et al. (2013). Activation of the PD-1 pathway contributes to immune escape in EGFR-driven lung tumors. *Cancer Discovery, 3*(12), 1355–1363.

Anderson, C. C. (2014). Application of central immunologic concepts to cancer: Helping T cells and B cells become intolerant of tumors. *European Journal of Immunology, 44*(7), 1921–1924.

Arai, H., Hori, S., Aramori, I., Ohkubo, H., & Nakanishi, S. (1990). Cloning and expression of a cDNA encoding an endothelin receptor. *Nature, 348*(6303), 730–732.

Assadipour, Y., Zacharakis, N., Crystal, J. S., Prickett, T. D., Gartner, J. J., Somerville, R. P. T., et al. (2017). Characterization of an immunogenic mutation in a patient with metastatic triple-negative breast cancer. *Clinical Cancer Research, 23*(15), 4347–4353.

Azuma, T., Yao, S., Zhu, G., Flies, A. S., Flies, S. J., & Chen, L. (2008). B7-H1 is a ubiquitous antiapoptotic receptor on cancer cells. *Blood, 111*(7), 3635–3643.

Bertolino, P., Trescol-Biemont, M. C., & Rabourdin-Combe, C. (1998). Hepatocytes induce functional activation of naive CD8+ T lymphocytes but fail to promote survival. *European Journal of Immunology, 28*(1), 221–236.

Bretscher, P., & Cohn, M. (1970). A theory of self-nonself discrimination. *Science (New York, NY), 169*(3950), 1042–1049.

de la Cruz-Merino, L., Henao Carrasco, F., Vicente Baz, D., Nogales Fernandez, E., Reina Zoilo, J. J., Codes Manuel de Villena, M., et al. (2011). Immune microenvironment in colorectal cancer: A new hallmark to change old paradigms. *Clinical & Developmental Immunology, 2011*, 174149.

Derbinski, J., Schulte, A., Kyewski, B., & Klein, L. (2001). Promiscuous gene expression in medullary thymic epithelial cells mirrors the peripheral self. *Nature Immunology, 2*(11), 1032–1039.

D'haeseleer, M., Beelen, R., Fierens, Y., Cambron, M., Vanbinst, A. M., Verborgh, C., et al. (2013). Cerebral hypoperfusion in multiple sclerosis is reversible and mediated by endothelin-1. *Proceedings of the National Academy of Sciences of the United States of America, 110*(14), 5654–5658.

Ding, H., Nedrud, J. G., Blanchard, T. G., Zagorski, B. M., Li, G., Shiu, J., et al. (2013). Th1-mediated immunity against Helicobacter pylori can compensate for lack of Th17 cells and can protect mice in the absence of immunization. *PLoS One, 8*(7), e69384.

Dunn, G. P., Bruce, A. T., Ikeda, H., Old, L. J., & Schreiber, R. D. (2002). Cancer immunoediting: From immunosurveillance to tumor escape. *Nature Immunology, 3*(11), 991–998.

Eil, R., Vodnala, S. K., Clever, D., Klebanoff, C. A., Sukumar, M., Pan, J. H., et al. (2016). Ionic immune suppression within the tumour microenvironment limits T cell effector function. *Nature, 537*(7621), 539–543.

Elisa, T., Antonio, P., Giuseppe, P., Alessandro, B., Giuseppe, A., Federico, C., et al. (2015). Endothelin receptors expressed by immune cells are involved in modulation of inflammation and in fibrosis: Relevance to the pathogenesis of systemic sclerosis. *Journal of Immunology Research, 2015*, 147616.

Gallucci, S., Lolkema, M., & Matzinger, P. (1999). Natural adjuvants: Endogenous activators of dendritic cells. *Nature Medicine, 5*(11), 1249–1255.

Goonetilleke, N., Liu, M. K., Salazar-Gonzalez, J. F., Ferrari, G., Giorgi, E., Ganusov, V. V., et al. (2009). The first T cell response to transmitted/founder virus contributes to the control of acute viremia in HIV-1 infection. *The Journal of Experimental Medicine, 206*(6), 1253–1272.

Gubin, M. M., Zhang, X., Schuster, H., Caron, E., Ward, J. P., Noguchi, T., et al. (2014). Checkpoint blockade cancer immunotherapy targets tumour-specific mutant antigens. *Nature, 515*(7528), 577–581.

Hammond, T. R., McEllin, B., Morton, P. D., Raymond, M., Dupree, J., & Gallo, V. (2015). Endothelin-B receptor activation in astrocytes regulates the rate of oligodendrocyte regeneration during remyelination. *Cell Reports, 13*(10), 2090–2097.

Haq, A., El-Ramahi, K., Al-Dalaan, A., & Al-Sedairy, S. T. (1999). Serum and synovial fluid concentrations of endothelin-1 in patients with rheumatoid arthritis. *Journal of Medicine, 30*(1–2), 51–60.

Hori, J., Wang, M., Miyashita, M., Tanemoto, K., Takahashi, H., Takemori, T., et al. (2006). B7-H1-induced apoptosis as a mechanism of immune privilege of corneal allografts. *Journal of immunology (Baltimore, MD: 1950), 177*(9), 5928–5935.

Hu, Y., Kim, H., Blackwell, C. M., & Slingluff, C. L., Jr. (2015). Long-term outcomes of helper peptide vaccination for metastatic melanoma. *Annals of Surgery, 262*(3), 456–464 [discussion 462–464].

Janeway, C. A., Jr. (1992). The immune system evolved to discriminate infectious nonself from noninfectious self. *Immunology Today, 13*(1), 11–16.

Jin, R., Aili, A., Wang, Y., Wu, J., Sun, X., Zhang, Y., et al. (2017). Critical role of SP thymocyte motility in regulation of thymic output in neonatal aire$-/-$ mice. *Oncotarget, 8*(1), 83–94.

Kantoff, P. W., Higano, C. S., Shore, N. D., Berger, E. R., Small, E. J., Penson, D. F., et al. (2010). Sipuleucel-T immunotherapy for castration-resistant prostate cancer. *The New England Journal of Medicine, 363*(5), 411–422.

Kauffman, H. M., McBride, M. A., & Delmonico, F. L. (2000). First report of the united network for organ sharing transplant tumor registry: Donors with a history of cancer. *Transplantation, 70*(12), 1747–1751.

Keir, M. E., Butte, M. J., Freeman, G. J., & Sharpe, A. H. (2008). PD-1 and its ligands in tolerance and immunity. *Annual Review of Immunology, 26*, 677–704.

Kelland, L. (2006). Discontinued drugs in 2005: Oncology drugs. *Expert Opinion on Investigational Drugs, 15*(11), 1309–1318.

Klein, L., Hinterberger, M., Wirnsberger, G., & Kyewski, B. (2009). Antigen presentation in the thymus for positive selection and central tolerance induction. *Nature Reviews Immunology, 9*(12), 833–844.

Kmieciak, M., Knutson, K. L., Dumur, C. I., & Manjili, M. H. (2007). HER-2/neu antigen loss and relapse of mammary carcinoma are actively induced by T cell-mediated antitumor immune responses. *European Journal of Immunology, 37*(3), 675–685.

Koebel, C. M., Vermi, W., Swann, J. B., Zerafa, N., Rodig, S. J., Old, L. J., et al. (2007). Adaptive immunity maintains occult cancer in an equilibrium state. *Nature, 450*(7171), 903–907.

Le, D. T., Lutz, E., Uram, J. N., Sugar, E. A., Onners, B., Solt, S., et al. (2013). Evaluation of ipilimumab in combination with allogeneic pancreatic tumor cells transfected with a GM-CSF gene in previously treated pancreatic cancer. *Journal of Immunotherapy (Hagerstown, MD), 36*(7), 382–389.

Li, J., Park, J., Foss, D., & Goldschneider, I. (2009). Thymus-homing peripheral dendritic cells constitute two of the three major subsets of dendritic cells in the steady-state thymus. *The Journal of Experimental Medicine, 206*(3), 607–622.

Liu, Y., Liang, X., Yin, X., Lv, J., Tang, K., Ma, J., et al. (2017). Blockade of IDO-kynurenine-AhR metabolic circuitry abrogates IFN-gamma-induced immunologic dormancy of tumor-repopulating cells. *Nature Communications, 8*, 15207.

Lu, H., Ladd, J., Feng, Z., Wu, M., Goodell, V., Pitteri, S. J., et al. (2012). Evaluation of known oncoantibodies, HER2, p53, and cyclin B1, in prediagnostic breast cancer sera. *Cancer Prevention Research (Philadelphia, PA), 5*(8), 1036–1043.

Malendowicz, L. K., Brelinska, R., De Caro, R., Trejer, M., & Nussdorfer, G. G. (1998). Endothelin-1, acting via the A receptor subtype, stimulates thymocyte proliferation in the rat. *Life Sciences, 62*(21), 1959–1963.

Manjili, M. H. (2014). The adaptation model of immunity. *Immunotherapy, 6*(1), 59–70.

Manjili, M. H. (2017). Tumor dormancy and relapse: From a natural byproduct of evolution to a disease state. *Cancer Research, 77*(10), 2564–2569.

Manjili, M. H., & Toor, A. A. (2014). Etiology of GVHD: Alloreactivity or impaired cellular adaptation? *Immunological Investigations, 43*(8), 851–857.

Martin, R., Whitaker, J. N., Rhame, L., Goodin, R. R., & McFarland, H. F. (1994). Citrulline-containing myelin basic protein is recognized by T-cell lines derived from multiple sclerosis patients and healthy individuals. *Neurology, 44*(1), 123–129.

Matzinger, P. (2002). The danger model: A renewed sense of self. *Science (New York, NY), 296*(5566), 301–305.

Matzinger, P. (2012). The evolution of the danger theory. Interview by Lauren Constable, commissioning editor. *Expert Review of Clinical Immunology, 8*(4), 311–317.

Michaud, M., Martins, I., Sukkurwala, A. Q., Adjemian, S., Ma, Y., Pellegatti, P., et al. (2011). Autophagy-dependent anticancer immune responses induced by chemotherapeutic agents in mice. *Science (New York, NY), 334*(6062), 1573–1577.

Michaud, M., Xie, X., Bravo-San Pedro, J. M., Zitvogel, L., White, E., & Kroemer, G. (2014). An autophagy-dependent anticancer immune response determines the efficacy of melanoma chemotherapy. *Oncoimmunology, 3*(7), e944047.

Mozaffarian, N., Wiedeman, A. E., & Stevens, A. M. (2008). Active systemic lupus erythematosus is associated with failure of antigen-presenting cells to express programmed death ligand-1. *Rheumatology (Oxford, England), 47*(9), 1335–1341.

Murata, S., Sasaki, K., Kishimoto, T., Niwa, S., Hayashi, H., Takahama, Y., et al. (2007). Regulation of CD8+ T cell development by thymus-specific proteasomes. *Science (New York, NY), 316*(5829), 1349–1353.

Myron Kauffman, H., McBride, M. A., Cherikh, W. S., Spain, P. C., Marks, W. H., & Roza, A. M. (2002). Transplant tumor registry: Donor related malignancies. *Transplantation, 74*(3), 358–362.

Nelson, J., Bagnato, A., Battistini, B., & Nisen, P. (2003). The endothelin axis: Emerging role in cancer. *Nature Reviews Cancer, 3*(2), 110–116.

Nelson, J. B., Chan-Tack, K., Hedican, S. P., Magnuson, S. R., Opgenorth, T. J., Bova, G. S., et al. (1996). Endothelin-1 production and decreased endothelin B receptor expression in advanced prostate cancer. *Cancer Research, 56*(4), 663–668.

Nelson, J. B., Udan, M. S., Guruli, G., & Pflug, B. R. (2005). Endothelin-1 inhibits apoptosis in prostate cancer. *Neoplasia (New York, NY), 7*(7), 631–637.

Odunsi, K., Matsuzaki, J., James, S. R., Mhawech-Fauceglia, P., Tsuji, T., Miller, A., et al. (2014). Epigenetic potentiation of NY-ESO-1 vaccine therapy in human ovarian cancer. *Cancer Immunology Research, 2*(1), 37–49.

Palmer, J. C., Barker, R., Kehoe, P. G., & Love, S. (2012). Endothelin-1 is elevated in Alzheimer's disease and upregulated by amyloid-beta. *Journal of Alzheimer's Disease, 29*(4), 853–861.

Parsa, A. T., Waldron, J. S., Panner, A., Crane, C. A., Parney, I. F., Barry, J. J., et al. (2007). Loss of tumor suppressor PTEN function increases B7-H1 expression and immunoresistance in glioma. *Nature Medicine, 13*(1), 84–88.

Paulsen, M., & Janssen, O. (2011). Pro- and anti-apoptotic CD95 signaling in T cells. *Cell Communication and Signaling, 9*, 7. https://doi.org/10.1186/1478-811X-9-7.

Payne, K. K., Keim, R. C., Graham, L., Idowu, M. O., Wan, W., Wang, X. Y., et al. (2016). Tumor-reactive immune cells protect against metastatic tumor and induce immunoediting of indolent but not quiescent tumor cells. *Journal of Leukocyte Biology*, *100*(3), 625–635.

Pulko, V., Harris, K. J., Liu, X., Gibbons, R. M., Harrington, S. M., Krco, C. J., et al. (2011). B7-h1 expressed by activated CD8 T cells is essential for their survival. *Journal of Immunology (Baltimore, MD)*, *187*(11), 5606–5614.

Rajasalu, T., Brosi, H., Schuster, C., Spyrantis, A., Boehm, B. O., Chen, L., et al. (2010). Deficiency in B7-H1 (PD-L1)/PD-1 coinhibition triggers pancreatic beta-cell destruction by insulin-specific, murine CD8 T-cells. *Diabetes*, *59*(8), 1966–1973.

Robert, C., Schachter, J., Long, G. V., Arance, A., Grob, J. J., Mortier, L., et al. (2015). Pembrolizumab versus ipilimumab in advanced melanoma. *The New England Journal of Medicine*, *372*(26), 2521–2532.

Santisteban, M., Reiman, J. M., Asiedu, M. K., Behrens, M. D., Nassar, A., Kalli, K. R., et al. (2009). Immune-induced epithelial to mesenchymal transition in vivo generates breast cancer stem cells. *Cancer Research*, *69*(7), 2887–2895.

Simonson, M. S. (1993). Endothelins: Multifunctional renal peptides. *Physiological Reviews*, *73*(2), 375–411.

Sistigu, A., Yamazaki, T., Vacchelli, E., Chaba, K., Enot, D. P., Adam, J., et al. (2014). Cancer cell-autonomous contribution of type I interferon signaling to the efficacy of chemotherapy. *Nature Medicine*, *20*(11), 1301–1309.

Slamon, D. J., Leyland-Jones, B., Shak, S., Fuchs, H., Paton, V., Bajamonde, A., et al. (2001). Use of chemotherapy plus a monoclonal antibody against HER2 for metastatic breast cancer that overexpresses HER2. *The New England Journal of Medicine*, *344*(11), 783–792.

Slingluff, C. L., Jr., Lee, S., Zhao, F., Chianese-Bullock, K. A., Olson, W. C., Butterfield, L. H., et al. (2013). A randomized phase II trial of multiepitope vaccination with melanoma peptides for cytotoxic T cells and helper T cells for patients with metastatic melanoma (E1602). *Clinical Cancer Research*, *19*(15), 4228–4238.

Snook, A. E., Magee, M. S., Schulz, S., & Waldman, S. A. (2014). Selective antigen-specific CD4(+) T-cell, but not CD8(+) T- or B-cell, tolerance corrupts cancer immunotherapy. *European Journal of Immunology*, *44*(7), 1956–1966.

Spirig, R., Potapova, I., Shaw-Boden, J., Tsui, J., Rieben, R., & Shaw, S. G. (2009). TLR2 and TLR4 agonists induce production of the vasoactive peptide endothelin-1 by human dendritic cells. *Molecular Immunology*, *46*(15), 3178–3182.

Steinaa, L., Rasmussen, P. B., Rygaard, J., Mouritsen, S., & Gautam, A. (2007). Generation of autoreactive CTL by tumour vaccines containing foreign T helper epitopes. *Scandinavian Journal of Immunology*, *65*(3), 240–248.

Swope, V. B., & Abdel-Malek, Z. A. (2016). Significance of the melanocortin 1 and endothelin B receptors in melanocyte homeostasis and prevention of sun-induced genotoxicity. *Frontiers in Genetics*, *7*, 146.

Takaba, H., Morishita, Y., Tomofuji, Y., Danks, L., Nitta, T., Komatsu, N., et al. (2015). Fezf2 orchestrates a thymic program of self-antigen expression for immune tolerance. *Cell*, *163*(4), 975–987.

Taube, J. M., Anders, R. A., Young, G. D., Xu, H., Sharma, R., McMiller, T. L., et al. (2012). Colocalization of inflammatory response with B7-h1 expression in human melanocytic lesions supports an adaptive resistance mechanism of immune escape. *Science Translational Medicine*, *4*(127). 127ra37.

Terracina, K. P., Graham, L. J., Payne, K. K., Manjili, M. H., Baek, A., Damle, S. R., et al. (2016). DNA methyltransferase inhibition increases efficacy of adoptive cellular immunotherapy of murine breast cancer. *Cancer Immunology, Immunotherapy*, *65*(9), 1061–1073.

Topalian, S. L., Hodi, F. S., Brahmer, J. R., Gettinger, S. N., Smith, D. C., McDermott, D. F., et al. (2012). Safety, activity, and immune correlates of anti-PD-1 antibody in cancer. *The New England Journal of Medicine, 366*(26), 2443–2454.

Topalian, S. L., Sznol, M., McDermott, D. F., Kluger, H. M., Carvajal, R. D., Sharfman, W. H., et al. (2014). Survival, durable tumor remission, and long-term safety in patients with advanced melanoma receiving nivolumab. *Journal of Clinical Oncology, 32*(10), 1020–1030.

Trabattoni, D., Saresella, M., Biasin, M., Boasso, A., Piacentini, L., Ferrante, P., et al. (2003). B7-H1 is up-regulated in HIV infection and is a novel surrogate marker of disease progression. *Blood, 101*(7), 2514–2520.

Tran, E., Ahmadzadeh, M., Lu, Y. C., Gros, A., Turcotte, S., Robbins, P. F., et al. (2015). Immunogenicity of somatic mutations in human gastrointestinal cancers. *Science (New York, NY), 350*(6266), 1387–1390.

Tran, E., Robbins, P. F., Lu, Y. C., Prickett, T. D., Gartner, J. J., Jia, L., et al. (2016). T-cell transfer therapy targeting mutant KRAS in cancer. *The New England Journal of Medicine, 375*(23), 2255–2262.

Tran, E., Turcotte, S., Gros, A., Robbins, P. F., Lu, Y. C., Dudley, M. E., et al. (2014). Cancer immunotherapy based on mutation-specific CD4+ T cells in a patient with epithelial cancer. *Science (New York, NY), 344*(6184), 641–645.

Verdegaal, E. M., de Miranda, N. F., Visser, M., Harryvan, T., van Buuren, M. M., Andersen, R. S., et al. (2016). Neoantigen landscape dynamics during human melanoma-T cell interactions. *Nature, 536*(7614), 91–95.

Wang, Q. J., Yu, Z., Griffith, K., Hanada, K., Restifo, N. P., & Yang, J. C. (2016). Identification of T-cell receptors targeting KRAS-mutated human tumors. *Cancer Immunology Research, 4*(3), 204–214.

Weed, D. T., Vella, J. L., Reis, I. M., De la Fuente, A. C., Gomez, C., Sargi, Z., et al. (2015). Tadalafil reduces myeloid-derived suppressor cells and regulatory T cells and promotes tumor immunity in patients with head and neck squamous cell carcinoma. *Clinical Cancer Research, 21*(1), 39–48.

Wulfing, P., Kersting, C., Tio, J., Fischer, R. J., Wulfing, C., Poremba, C., et al. (2004). Endothelin-1-, endothelin-A-, and endothelin-B-receptor expression is correlated with vascular endothelial growth factor expression and angiogenesis in breast cancer. *Clinical Cancer Research, 10*(7), 2393–2400.

Yu, L., Wang, L., & Chen, S. (2010). Endogenous toll-like receptor ligands and their biological significance. *Journal of Cellular and Molecular Medicine, 14*(11), 2592–2603.

Yuen, T. J., Johnson, K. R., Miron, V. E., Zhao, C., Quandt, J., Harrisingh, M. C., et al. (2013). Identification of endothelin 2 as an inflammatory factor that promotes central nervous system remyelination. *Brain, 136*(Pt 4), 1035–1047.

Ziegler, A., Muller, C. A., Bockmann, R. A., & Uchanska-Ziegler, B. (2009). Low-affinity peptides and T-cell selection. *Trends in Immunology, 30*(2), 53–60.

CHAPTER THREE

# Bcl-2 Antiapoptotic Family Proteins and Chemoresistance in Cancer

Santanu Maji*[,†], Sanjay Panda[‡,§], Sabindra K. Samal*[,†],
Omprakash Shriwas*[,†], Rachna Rath[¶], Maurizio Pellecchia[∥],
Luni Emdad[#,**,††], Swadesh K. Das[#,**,††], Paul B. Fisher[#,**,††],
Rupesh Dash*[,1]

*Institute of Life Sciences, Bhubaneswar, Odisha, India
†Manipal University, Manipal, Karnataka, India
‡HCG Panda Cancer Centre, Cuttack, Odisha, India
§Acharya Harihar Regional Cancer Centre, Cuttack, Odisha, India
¶Sriram Chandra Bhanj Dental College and Hospital, Cuttack, Odisha, India
∥University of California, Riverside, Riverside, CA, United States
#Virginia Commonwealth University, School of Medicine, Richmond, VA, United States
**VCU Institute of Molecular Medicine, Virginia Commonwealth University, School of Medicine, Richmond, VA, United States
††VCU Massey Cancer Center, Virginia Commonwealth University, School of Medicine, Richmond, VA, United States
[1]Corresponding author: e-mail address: rupesh.dash@gmail.com

## Contents

1. Introduction ............................................. 38
   1.1 Chemotherapy: A Standard Treatment for Cancer ........ 38
   1.2 Chemoresistance ..................................... 39
2. Bcl-2 Family Proteins and Chemoresistance ............... 49
   2.1 Bcl-2 Family Proteins and Apoptosis .................. 49
   2.2 Role of Bcl-2 in Chemoresistance ..................... 50
   2.3 Role of Mcl-1 in Chemoresistance ..................... 52
   2.4 Role of Bcl-xL in Chemoresistance .................... 53
   2.5 Role of Bfl-1/A1 in Chemoresistance .................. 54
3. Potential Bcl-2 Inhibitors in Chemoresistance: Clinical Development ... 55
   3.1 ABT-263 (Navitoclax) ................................. 55
   3.2 ABT-199 (Venetoclax) ................................. 55
   3.3 AT-101 (Gossypol) .................................... 55
   3.4 GX15-070 (Obatoclax) ................................. 62
   3.5 Oblimersen .......................................... 62
4. Conclusions and Future Perspectives ..................... 63
Acknowledgments ............................................ 63
References ................................................. 64

## Abstract

Cancer is a daunting global problem confronting the world's population. The most frequent therapeutic approaches include surgery, chemotherapy, radiotherapy, and more recently immunotherapy. In the case of chemotherapy, patients ultimately develop resistance to both single and multiple chemotherapeutic agents, which can culminate in metastatic disease which is a major cause of patient death from solid tumors. Chemoresistance, a primary cause of treatment failure, is attributed to multiple factors including decreased drug accumulation, reduced drug–target interactions, increased populations of cancer stem cells, enhanced autophagy activity, and reduced apoptosis in cancer cells. Reprogramming tumor cells to undergo drug-induced apoptosis provides a promising and powerful strategy for treating resistant and recurrent neoplastic diseases. This can be achieved by downregulating dysregulated antiapoptotic factors or activation of proapoptotic factors in tumor cells. A major target of dysregulation in cancer cells that can occur during chemoresistance involves altered expression of *Bcl-2* family members. Bcl-2 antiapoptotic molecules (Bcl-2, Bcl-xL, and Mcl-1) are frequently upregulated in acquired chemoresistant cancer cells, which block drug-induced apoptosis. We presently overview the potential role of Bcl-2 antiapoptotic proteins in the development of cancer chemoresistance and overview the clinical approaches that use Bcl-2 inhibitors to restore cell death in chemoresistant and recurrent tumors.

## 1. INTRODUCTION

Cancer remains a major global public health concern. According to the CDC (Centers for Disease Control and Prevention), each year about 14.1 million new cases of cancer are diagnosed and 8.2 million lives are lost due to cancer. In 2017, approximately 688,780 new cancer cases and 600,920 cancer deaths are projected in the United States alone (Siegel, Miller, & Jemal, 2017). From its original discovery, multiple treatment modalities have been adopted to treat patients with cancer. Although there have been breakthroughs and successes in treating specific types of cancer, the majority of strategies have not proven as efficacious as hoped or predicted. Despite this lack of optimum success, chemotherapy still remains as one of the primary treatment modalities for cancer, either used alone or in combination with other therapeutic approaches.

### 1.1 Chemotherapy: A Standard Treatment for Cancer

Chemotherapy was first used clinically in 1940 with the application of nitrogen mustards, and it remains a foundation of clinical practice for cancer treatment. In 1942, nitrogen mustards were shown to inhibit lymphoid tumor growth in a patient with non-Hodgkin's lymphoma (NHL) (Gilman, 1963). In the late 1940s, the anticancer effect of folic acid was established in AML patients (Chabner & Roberts, 2005). In 1958, it was found that folate analog amethopterin (methotrexate) suppressed choriocarcinomas (Li, Hertz, & Bergenstal, 1958). For decades, methotrexate was not only used as an effective chemotherapy for leukemia and lymphomas, but it also has been successfully used in solid tumors like lung, breast, head and neck, bladder, and gestational trophoblastic carcinomas (Li et al., 1958; Skubisz & Tong, 2012). Purine analogs, such as 6-mercaptopurine (6-MP), have been used as anticancer agents because they block de novo DNA and RNA synthesis (Hitchings & Elion, 1954; Skipper, Thomson, Elion, & Hitchings, 1954). The Eli Lilly pharmaceutical group demonstrated that an antidiabetic agent, vinca alkaloid, could significantly inhibit tumor cell proliferation (Johnson, Armstrong, Gorman, & Burnett, 1963). In 1965, methotrexate, vinca alkaloid, 6-MP, and prednisone were successfully used as a combinatorial therapy in children with acute lymphocytic leukemia (ALL) (Frei et al., 1965). Although cisplatin (*cis*-dichlorodiammineplatinum) was synthesized in 1844 by M. Peyrone, it was first evaluated and reported by Rosenberg in 1960 as a bacterial growth inhibitor (Rosenberg, Vancamp, & Krigas, 1965). Since then, platinum-based drugs have been widely employed as chemotherapy for various neoplasms. Cisplatin and its derivative carboplatin are effectively used as chemotherapy for ovarian, lung, and head and neck cancers (Go & Adjei, 1999). In combination, cisplatin and 5-fluorouracil (5-FU) block head and neck, breast, small lung, and ovarian carcinoma growth (Kish et al., 1982; Klaassen et al., 1997; Kucuk, Shevrin, Pandya, & Bonomi, 2000; Morgan et al., 2000). Similarly, a combination of cisplatin and paclitaxel has significant anticancer effects in ovarian, breast, non-small-cell lung, and head and neck carcinomas (de Souza Viana et al., 2016; du Bois et al., 2003; Rosell et al., 2002; Wasserheit et al., 1996). Docetaxel and carboplatin treatment showed a substantial response in taxane nonresponding prostate cancer patients (Oh, George, & Tay, 2005). Imatinib (an inhibitor of the BCR–ABL tyrosine kinase) has shown potential for the chemotherapy of leukemia (Iqbal & Iqbal, 2014).

## 1.2 Chemoresistance

Despite initial positive responses with chemotherapy, many cancer patients experience relapse and continued tumor growth and spread due to drug resistance, which leads to treatment failure and metastatic disease. Moreover, chemoresistance is a major contributing factor that reduces the efficacy of drug treatment. The conventional chemotherapy drugs efficiently eliminate the rapidly dividing cancer cells by inducing cell death, but poorly target slowly dividing cells or disseminated tumor cells (Blagosklonny, 2006; Linde, Fluegen, & Aguirre-Ghiso, 2016). In addition, in many cases rapidly dividing cells do not respond to chemotherapy, particularly when a low dose is provided to offset adverse effects on normal cells. These poorly sensitive cancer cells or populations of cancer cells ultimately contribute to tumor recurrence. Chemoresistance is broadly divided into two types: "intrinsic chemoresistance" where cancer cells are resistant prior to chemotherapy and "acquired chemoresistance" where cancer cells develop resistance during prolonged treatment with agents that initially displayed sensitivity (Kerbel, Kobayashi, & Graham, 1994). In addition to this, while acquiring chemoresistance against a particular chemotherapeutic drug, the tumor may acquire cross-resistance to a range of alternative drugs resulting in development of multidrug resistance (MDR). For decades, researchers have tried to understand the molecular mechanism of chemoresistance in various cancer neoplasms. Multiple genetic and epigenetic factors and pathways can contribute to resistance to chemotherapy, which is summarized schematically in Fig. 1 and briefly discussed below.

### 1.2.1 Cancer Stem Cell

It is hypothesized that tumors consist of a heterogeneous population of cells and a small subpopulation of cells known as cancer-initiating cells or cancer stem cells (CSCs). CSCs are not only resistant to chemotherapy but also have

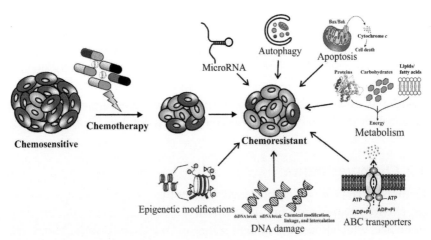

**Fig. 1** Various factors involved in chemoresistance.

enhanced tumor-initiating abilities, which contribute to chemoresistance and recurrence (Guo, Lasky, & Wu, 2006). Despite intensive investigations to define genetic and pharmacological approaches to target CSCs, they remain a significant clinical challenge in overcoming cancer chemoresistance (Talukdar, Emdad, Das, Sarkar, & Fisher, 2016). Unfortunately, conventional chemotherapeutics are frequently not effective in targeting CSCs, but rather enrich for CSC population by selecting the clones which are resistant to drug treatment. CSCs contribute to acquired chemoresistance in several neoplasms including ovarian cancer (Deng et al., 2016; Steg et al., 2012), breast cancer (Abdullah & Chow, 2013; Saha et al., 2016; Vidal, Rodriguez-Bravo, Galsky, Cordon-Cardo, & Domingo-Domenech, 2014), prostate cancer (Mayer, Klotz, & Venkateswaran, 2015; Ni et al., 2014), lung cancer (Ham et al., 2016; Hsu et al., 2011; Yang et al., 2016; Zhang et al., 2017), colorectal cancer (Bose et al., 2011; Garza-Trevino, Said-Fernandez, & Martinez-Rodriguez, 2015; Hu et al., 2015), and oral cancer (Chen, Wu, et al., 2017; Gao et al., 2017; Ghuwalewala et al., 2016). Accordingly, understanding the molecular mechanism of CSC involvement in chemoresistance remains a major concern for researchers and clinicians. Current studies in prostate cancer showed that $CD117^+/ABCG2^+$ subpopulations of cells have enhanced expression of stem cell markers such as Nanog, Oct4, Sox2, Nestin, and CD133, and these populations are resistant to chemotherapeutics such as cisplatin, paclitaxel, adriamycin, and methotrexate (Liu et al., 2010). Also, in non-small-cell lung cancer (NSCLC) cells, cisplatin induces CSCs via enhancing TRIB1/HDAC activity, and these cells contribute to MDR (Wang et al., 2017). In breast cancer, CSCs contribute to cisplatin resistance in Brca1/p53-mediated mouse mammary tumors (Shafee et al., 2008). Overexpression of Mucin-1 enhances CSC markers and sphere formation ability in paclitaxel-resistant lung cancer cells (Ham et al., 2016). Abundant expression of ABC transporters in CSCs helps in drug efflux and enhances the development of chemoresistance (Dean, 2009; Moitra, 2015; Talukdar et al., 2016). Looking at the important contribution of CSCs in acquiring chemoresistance, several clinical trials have been conducted with various compounds that target CSCs in combination with conventional chemotherapeutic agents. These compounds either induce CSC death or induce CSC differentiation by inhibiting important stemness pathways such as Wnt, Notch, and Hedgehog.

### 1.2.2 Upregulation of ABC Transporter Proteins

Increased drug efflux is one of the hallmarks of chemoresistance, and a major contributor to this phenotype is overexpression of ATP-binding cassette

(ABC) transporters (Gottesman, 2002; Wu, Calcagno, & Ambudkar, 2008). The ABC transporters are primarily responsible for transport of ions and xenobiotic drugs through the cell membranes using ATPase-based channel proteins. The structure of ABC transporters includes minimum two transmembrane (TM) domains and two nucleotide domains. The TM domains recognize and translocate the ions and drugs, and the nucleotide domain has an ATP-binding site. Three different types of ABC transporters are frequently upregulated and broadly linked with both intrinsic and acquired chemoresistance in various neoplasms, i.e., multidrug resistance gene R1 MDR1 (ABCB1) also known as P-glycoprotein (Linn & Giaccone, 1995; Mahon et al., 2003; Schneider et al., 2001; Zhou et al., 2014), MDR-associated protein 1 (MRP1 or ABCC1) (Cai et al., 2011; Liu, Li, et al., 2014; Triller, Korosec, Kern, Kosnik, & Debeljak, 2006; Zalcberg et al., 2000), and breast cancer resistance protein (BCRP or ABCG2) (Gao, Zhang, Wang, & Ren, 2016; He et al., 2016; Sabnis, Miller, Titus, & Huss, 2017; Stacy, Jansson, & Richardson, 2013; Zhao, Ren, et al., 2015). In addition to these ABC transporters, ABCC3 also contributes to enhanced drug efflux in breast, non-small-cell lung, and colon carcinomas (Balaji, Udupa, Chamallamudi, Gupta, & Rangarajan, 2016; Jiang et al., 2009; Zhao et al., 2013). In another example, ACC10 (MRP7) is overexpressed in paclitaxel-resistant NSCLC cells, and downregulation of ACC10 expression sensitizes these cells to paclitaxel (Sun et al., 2013). Several approaches have been adopted to overcome ABC transporter-mediated chemoresistance in cancer. These include (i) developing small-molecule inhibitors for ABC transporters like elacridar, laniquidar, and zosuquidar; (ii) use of tyrosine kinase inhibitors as ABC transporter regulators; (iii) targeting important tumorigenic pathways to inhibit ABC transporters; and (iv) nanoparticle-based chemotherapy drug delivery that bypasses the ABC transporter-mediated efflux (McIntosh, Balch, & Tiwari, 2016).

### 1.2.3 DNA Repair Mechanisms

DNA damage repair (DDR) mechanisms are essential for the survival of normal cells, which helps to maintain their genetic integrity against the stress induced by genotoxic agents. DNA lesions are sensed by DDR factors, which further trigger cell cycle check points followed by DNA repair. In response to chemotherapy, cancer cells develop defective repair systems that alter their sensitivity to chemotherapeutic drugs (Bartek, Bartkova, & Lukas, 2007; Wang, Mosel, Oakley, & Peng, 2012). Various DDR proteins are frequently deregulated in chemoresistant cancer cells, offsetting the

DNA-damaging properties of chemotherapy. Knocking down the DNA repair protein APE1/Ref-1 (APE1) in cisplatin-resistant melanomas enhances their sensitivity toward chemotherapeutic agents (Yang, Irani, Heffron, Jurnak, & Meyskens, 2005). An in vitro study described the expression of several DDR factors, including ATM, Mre11, and H2AX, which are significantly reduced, and DDR signaling is moderately malfunctioned in oral/laryngeal SCC cells causing chemoresistance against cisplatin (Wang et al., 2012). Preclinical models suggested that targeting poly(ADP-ribose) polymerase-1 by pharmacological inhibitors and alkylators exhibited a synergistic effect in sensitizing BRACA-deficient cancers (Calabrese et al., 2004; Donawho et al., 2007; Evers et al., 2008). Similarly, knocking down base excision repair proteins like $N$-methylpurine-DNA glycosylase or apurinic–apyrimidinic endonuclease 1 (APE1) sensitizes cancer cells to alkylating chemotherapeutics (Adhikari et al., 2008). Error-prone translational DNA synthesis (TLS) plays an important role in acquiring chemoresistance. In a preclinical study of B-cell lymphoma, it was found that suppression of Rev 1 (a TLS scaffold protein) reduced tumor drug resistance against cyclophosphamide (Xie, Doles, Hemann, & Walker, 2010). In another preclinical model of lung adenocarcinomas, it is found that the DNA pol$\zeta$ plays an integral role in the cisplatin resistance, and knocking down the expression of Rev3 (an essential component of pol$\zeta$) sensitizes drug-resistant tumor cells to chemotherapy (Doles et al., 2010). Based on the above-mentioned reports it can be suggested that the interactions between DDR pathways and related factors in response to chemotherapy may need a combinatorial approach for developing an efficient and effective chemotherapeutic strategy. DDR is now considered as a therapeutic target for chemoresistant cancers with inherent DNA repair deficiencies.

### *1.2.4 Epigenetic Effects*

Epigenetic changes involve modifications in gene expression without structurally altering genetic sequences (Berger, Kouzarides, Shiekhattar, & Shilatifard, 2009; Crea et al., 2011). Recently, several evidences suggested that modifications in the epigenetic landscape can reprogram cancer cells to acquire chemoresistance (Cacan, 2017; Choi et al., 2017; Dalvi et al., 2017; Fujita et al., 2015; Liu, Siu, et al., 2014; Zhao, Cao, et al., 2015). Major epigenetic modification affecting cancer cells includes DNA methylation and histone modification that alter the sensitivity to chemotherapy drugs (Ronnekleiv-Kelly, Sharma, & Ahuja, 2017). Hypermethylation of Notch3 causes activation of P-glycoprotein and modulates the sensitivity of adriamycin- and paclitaxel-resistant breast cancer cells (Gu et al., 2016). Similarly, DNA methylation of Dickkopf-related protein 3 modulates NSCLC cells via increased expression of P-glycoprotein, resulting in resistance toward docetaxel (Tao, Huang, Chen, & Chen, 2015). Methylation in tumor suppressors, BLU and RUNX3, results in acquired resistance against paclitaxel and docetaxel in ovarian and lung carcinomas (Chiang et al., 2013; Zhang et al., 2012). Class I and Class II histone deacetylases (HDACs) are involved in epigenetic alterations of several target genes. HDAC1 was found to be upregulated in cisplatin-resistant HNSCC, and suppressing the activity of HDAC1 by SAHA (suberoylanilide hydroxamic acid) reverses the cisplatin resistance (Kumar, Yadav, Lang, Teknos, & Kumar, 2015). In another study on biliary tract cancer, the HDAC inhibitor vorinostat regulates TGF-β1-mediated epithelial-to-mesenchymal transition (EMT) and sensitizes gemcitabine-resistant tumors (Sakamoto et al., 2016). Similarly, the histone deacetylase SIRT6 expression has been correlated with chemoresistance in NSCLC patients, and in vitro knockdown of SIRT6 enhanced paclitaxel sensitivity (Azuma et al., 2015). Furthermore, activation of SIRT1 was found in drug-resistant ovarian cancer, and knockdown of SIRT1 expression by siRNA sensitizes these cancer cells by obstructing the inhibitory effects of isoproterenol on doxorubicin-induced p53 acetylation (Chen, Zhang, Cheng, et al., 2017).

### 1.2.5 Autophagy-Associated Chemoresistance

Autophagy is a catabolic process which maintains the cell homeostasis by removing unnecessary or dysfunctional components. It is a process in which the cell organelles, potion of cytosol, and biomacromolecules are sequestered to autophagosomes and delivered to lysosomes for bulk degradation. Autophagy plays dual roles in tumorigenesis; it either induces tumor cell death or maintains cancer cell survival (Bhutia et al., 2013; Shintani & Klionsky, 2004). Current evidence suggests that during chemotherapy, autophagy is induced as a protective mechanism against stress to promote chemoresistance in various cancer neoplasms (Liu & Debnath, 2016; Sui et al., 2013). In breast cancer cells, autophagy induced by epirubicin and tamoxifen mediates chemoresistance, and blocking autophagy significantly restores drug-mediated cell death (Schoenlein, Periyasamy-Thandavan, Samaddar, Jackson, & Barrett, 2009). Suppression of autophagy resensitizes colorectal cancer cells to widely used chemotherapeutics, i.e., 5-FU and oxaliplatin (de la Cruz-Morcillo et al., 2012; Sasaki et al., 2010). In NSCLC patients, treatment with the autophagy inhibitor chloroquine enhances the cytotoxicity of EGFR tyrosine kinase inhibitors like gefitinib or erlotinib (Goldberg et al., 2012; Han et al., 2011). Genetic or pharmacological inhibition of autophagy significantly enhances ABT-737 toxicity in prostate cancer cells (Saleem et al., 2012). Similarly, clinical trials are being conducted to evaluate the combinatorial efficacy of IL-2 and chloroquine in patients having renal cell carcinomas (NCT01144169 and NCT01550367). In ovarian carcinomas, nucleus accumbens-1-induced autophagy mediates cisplatin resistance (Zhang et al., 2010). In the case of chemoresistance, the induction of autophagy is mediated through dysregulation of PI3K/AKT/mTOR, the known master regulators of autophagy. Inhibition of PI3K/mTOR pathway results in suppression of autophagy which restores cell death in drug-resistant cancer cells (Li, Jin, Zhang, Xing, & Kong, 2013). In addition, EGFR, VEGF, p53, and MAP kinase signaling also play important roles in chemotherapy-induced autophagy (Stanton et al., 2013). Considering the cytoprotective role of autophagy in drug-induced stress in cancer cells, it may provide a therapeutic target to overcome chemoresistance, which is evident from several clinical trials which include autophagy inhibitors along with the conventional chemotherapeutics for chemoresistant diseases (Galluzzi, Bravo-San Pedro, Levine, Green, & Kroemer, 2017).

### 1.2.6 Imbalances of MicroRNA

MicroRNAs (miRNAs) are short nonprotein-coding RNAs which inhibit gene expression at a posttranscriptional level (Filipowicz, Bhattacharyya, & Sonenberg, 2008). Chemoresistance in various cancer neoplasms involves imbalances in miRNA profiling (Croce & Fisher, 2017; Okamoto, Miyoshi, & Murawaki, 2013). In lung cancer cells, selective upregulation of miRNA-103, miRNA-203, and miRNA-21 promotes chemoresistance by targeting PKC, SRC, and caspase-8, respectively (Garofalo et al., 2011; Jeon et al., 2015). Similarly, in cholangiocarcinoma, downregulation of miR-221 and miR-29b induces the expression of PIK3R1 (phosphoinositide-3-kinase regulatory subunit 1) which mediates gemcitabine resistance (Okamoto et al., 2013). Significant downregulation of let-7i expression was found in ovarian chemotherapy-resistant patients ($n=69$, $P=0.003$), and this change associated with shorter progression-free survival (Yang et al., 2008). A recent study revealed that overexpression of miR-1307 promotes taxol resistance in ovarian cells by targeting the inhibitor of growth family protein-5 expression (Chen, Yang, et al., 2017). In colorectal cancer cells, miR-140 and miR-215 were found to be overexpressed and contribute to methotrexate, 5-FU, and tomudex chemoresistance

(Song et al., 2010, 2009). In ALDHA1$^+$ colorectal CSCs, overexpression of miR-199a/b leads to cisplatin resistance by activation of Wnt signaling pathway and upregulation of ABCG2 (Chen, Zhang, Kuai, et al., 2017). In paclitaxel-resistant colorectal cancer cells, ectopic expression of miR-203 overcomes drug resistance by targeting the salt-inducible kinase 2 (Liu et al., 2016). MicroRNA-30c contributes to breast cancer chemoresistance by regulating the actin-binding protein twinfilin 1 (Bockhorn et al., 2013). Understanding the molecular mechanism behind miRNA-mediated chemoresistance development is important for defining approaches for overcoming chemoresistance.

### 1.2.7 Altered Metabolic Pathways

Cancer cells have altered metabolic properties as compared to the normal cells, which enhance their survival. In addition to their "addiction" toward aerobic glycolysis, cancer cells also exhibit increased fatty acid synthesis and increased rates of glutamine metabolism. Several pieces of evidence indicate that dysregulated metabolism is linked to chemoresistance in cancer (Munoz-Pinedo, El Mjiyad, & Ricci, 2012). Enhanced glycolytic activity contributes to chemoresistance against glucocorticoids in childhood ALL, and inhibition of glycolysis by pharmacological inhibitors restores prednisolone-induced cell death (Hulleman et al., 2009). Similarly, PKM2 (an isoform of pyruvate kinase 2) negatively correlates with drug response against oxaliplatin in colorectal cancer (Munoz-Pinedo et al., 2012). Glucose transporters are frequently upregulated in chemoresistant cancers. In vitro studies indicate that targeting GLUT1 significantly enhances daunorubicin-, cisplatin-, and paclitaxel-induced toxicity in various cancer neoplasms (Cao et al., 2007; Liu et al., 2012). Similarly, GLUT3 was found to be upregulated in glioblastoma, thereby decreasing the efficiency of temozolomide therapy (Rahman & Hasan, 2015). Targeting GLUT4 by its inhibitor ritonavir increases sensitivity toward doxorubicin in multiple myeloma (McBrayer et al., 2012). Similarly, inhibiting the activity of the glycolytic enzyme hexokinase results in sensitizing ABT-737/ABT-263 in leukemia, cervical, breast, and prostate carcinomas (Coloff et al., 2011; Meynet et al., 2012; Yamaguchi et al., 2011). Lactate dehydrogenase-A (LDHA), an isoform of LDH, plays a key role in glucose metabolism and is responsible for taxol and trastuzumab resistance in breast cancer (Zhao et al., 2011; Zhou et al., 2010). Lipid biosynthesis and amino acid metabolism are also involved in acquiring drug resistance. FAS (fatty acid synthase), a crucial enzyme for lipid biosynthesis, is overexpressed in many cancer cells (Flavin, Peluso, Nguyen, & Loda, 2010). In breast and pancreatic carcinomas, it plays an important role in the development of chemoresistance against trastuzumab, adriamycin, docetaxel, 5-FU, and gemcitabine (Liu, Liu, & Zhang, 2008; Menendez, Lupu, & Colomer, 2004; Menendez, Vellon, & Lupu, 2005; Munoz-Pinedo et al., 2012; Vazquez-Martin, Ropero, Brunet, Colomer, & Menendez, 2007; Yang et al., 2011). This evidence suggests that important metabolic enzymes can serve as therapeutic targets for chemoresistant cancer.

### 1.2.8 Reduced Apoptotic Response in Chemoresistance

One important cell death mechanism is apoptosis (programmed cell death), which is essential for performing normal physiological functions and maintenance of organism homeostasis. Two distinct unrelated pathways can lead to apoptosis, the extrinsic/cell death receptor pathway and the intrinsic/mitochondrial pathway (Quinn et al., 2011; Thomas et al., 2013) (Fig. 2). The extrinsic apoptotic pathway is mediated by activation of death receptors belonging to the tumor necrosis factor receptor superfamily, such as FasL and TNFα. Activation of any of the death receptors results in the cleavage and activation of caspase-8, culminating in a signaling cascade called "death-inducing signaling complex," which leads to activation of caspase-3, ultimately inducing cell death (Ashkenazi & Dixit, 1999; Kischkel et al., 1995). The intrinsic apoptotic pathway is generally induced by a variety of stress signals including diverse cytotoxic events and involves mitochondrial outer membrane permeabilization, which leads to cytochrome $c$ release. Once released, cytochrome $c$ binds to Apaf-1 and forms the "apoptosome" that results in the cleavage and activation of caspase-9 followed by cleavage of caspase-3, ultimately causing cell death (Green & Kroemer, 2004; Kang & Reynolds, 2009). Almost all of the clinically used conventional chemotherapeutic drugs induce apoptosis in cancer cells. Inhibition of apoptosis in response to chemotherapeutic drugs is one of the central events during development of chemoresistance in cancer cells. The intrinsic apoptosis is majorly regulated by B-cell lymphoma-2 (Bcl-2) family proteins. This is attributed mostly to the altered expression pattern of antiapoptotic and proapoptotic proteins in chemoresistant cells. In chemoresistant cells, the Bcl-2 antiapoptotic proteins are frequently upregulated, offsetting the function of proapoptotic proteins. In this chapter, we discuss the role of Bcl-2 antiapoptotic family proteins in the development of chemoresistance and clinical applications of antiapoptotic inhibitors in recurrent diseases.

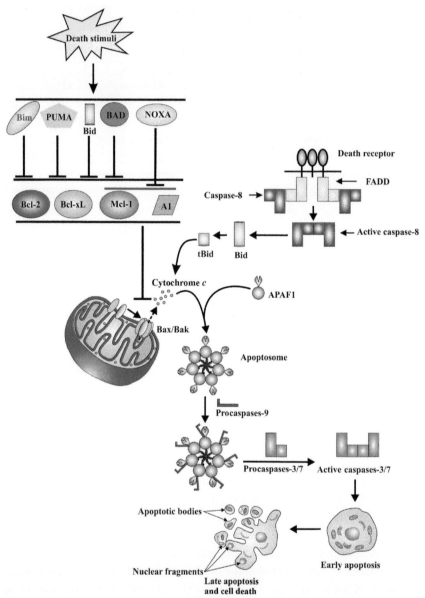

**Fig. 2** Pathways of programmed cell death and its involvement with Bcl-2 family proteins.

## 2. Bcl-2 FAMILY PROTEINS AND CHEMORESISTANCE
### 2.1 Bcl-2 Family Proteins and Apoptosis

The Bcl-2 family member proteins tightly regulate the intrinsic apoptotic pathway. Bcl-2 was originally identified as an oncogene, which was activated via chromosome translocation in human follicular lymphoma (Bakhshi et al., 1985). At the present time, about 25 known Bcl-2 family proteins have been discovered, and these proteins share certain sequence homologies through the presence of Bcl-2 homology (BH) domains. There are four different BH domains in the Bcl-2 family proteins, and each protein has at least one of these domains. Based on their functional activity, the Bcl-2 family is broadly divided into two major groups: (1) proapoptotic and (2) antiapoptotic Bcl-2-like proteins. The proapoptotic Bcl-2 family proteins are further classified into two subgroups: the multidomain effector proteins and the BH3-only proteins (Youle & Strasser, 2008). The multidomain effector proteins have all four BH domains, and these members include Bax (Bcl-2-associated x), Bak (Bcl-2 homologous agonist killer), and Bok (Bcl-2-related ovarian killer). The BH3-only proteins share sequence similarity with the rest of the family only through their BH3 domain. These Bcl-2 members include NOXA, PUMA, Bim, Bik, and Bid (Elkholi, Floros, & Chipuk, 2011). MOMP occurs only when Bax and Bak form dimers, i.e., homo/heterodimers. But in the normal scenario, the antiapoptotic proteins bind to proapoptotic effector proteins and prevent their dimerization. The BH3-only proteins compete and bind to antiapoptotic proteins thereby displacing Bax and Bak, allowing the free Bax and Bak to form dimers resulting in MOMP and release of cytochrome $c$ into the cytoplasm (Fig. 2).

The antiapoptotic Bcl-2 family proteins include Bcl-2, Bcl-xL (Bcl-2-related gene long isoform), Bcl-w, Mcl-1 (myeloid cell leukemia cell differentiation protein-1), and the Bcl-2-related gene A1. All these antiapoptotic proteins possess four BH domains and promote cell survival or inhibit apoptosis by inactivating their proapoptotic Bcl-2 family counterparts (Danial, 2007). Dysregulation of the Bcl-2 antiapoptotic protein is a common phenomenon during carcinogenesis. Transgenic mice overexpressing Bcl-2 develop spontaneous tumors (McDonnell & Korsmeyer, 1991), and upregulated Bcl-2 occurs in various neoplasms including breast carcinomas, prostate carcinomas, glioblastomas, and lymphomas (Placzek et al., 2010; Strik et al., 1999). Similarly, to evade cell death cancer cells exploit

Mcl-1 overexpression. Several investigators including ourselves have reported that Mcl-1 overexpression can also be found in several malignancies including prostate carcinomas, leukemias, pancreatic cancers, and oral cancers (Dash et al., 2011, 2010; Maji et al., 2015; Placzek et al., 2010) and hepatocellular carcinomas (Sieghart et al., 2006). In addition, tumor cells have also been found to acquire resistance to cancer therapeutics through overexpression of Bcl-2 antiapoptotic members (Campbell et al., 2010; Miyashita & Reed, 1993; Zhou, Qian, Kozopas, & Craig, 1997). Apart from the full-length Mcl-1 long form (Mcl-1L), the human Mcl-1 gene undergoes differential splicing, which yields Mcl-1 short (Mcl-1S) and Mcl-1 extra short (Mcl-1ES) splice variants. Whereas full-length Mcl-1 derives from three coding exons, the short splicing variant results from the deletion of 248 nucleotides (exon 2) from the full-length Mcl-1L cDNA. A shift in the open reading frame in Mcl-1S leads to the complete loss of BH1, BH2, and TM domains. Unlike Mcl-1L, Mcl-1S induces apoptosis upon ectopic expression in Chinese hamster ovary cells, and it dimerizes with Mcl-1L (Bae, Leo, Hsu, & Hsueh, 2000). Compared to the longest Mcl-1 the Mcl-1ES transcript has a truncated exon 1, resulting in loss of the PEST motifs but retention of the BH1 to BH3 and the TM domains (Kim, Sim, et al., 2009). Mcl-1ES contains three mutations at 242G, 501A, and 587A, encoding amino acids 81R, 167Q, and 196K, respectively. Ectopic expression of Mcl-1 ES moderately decreases cell viability in HeLa cells, but interestingly cell viability is significantly decreased when coexpressed with Mcl-1L (Kim, Sim, et al., 2009).

## 2.2 Role of Bcl-2 in Chemoresistance

Overexpression of Bcl-2 was correlated with poor clinical outcome in multiple neoplasms including AML, non-Hodgkin lymphoma, melanoma, breast cancer, and prostate cancer (Campos et al., 1993; Grover & Wilson, 1996; Hermine et al., 1996; Joensuu, Pylkkanen, & Toikkanen, 1994; McDonnell et al., 1992). In an *Em-myc* transgenic mouse model of lymphomas, studies indicated that overexpression of Bcl-2 inhibited adriamycin-, mafosphamide-, and docetaxel-induced apoptosis (Schmitt & Lowe, 2001). In gastric cancer, enhanced Bcl-2 expression was significantly correlated with chemoresistance to 5-FU ($r_s = 0.265$, $P = 0.041$), adriamycin ($r_s = 0.425$, $P = 0.001$), and mitomycin ($r_s = 0.40$, $P = 0.002$) (Geng, Wang, & Li, 2013). Cisplatin resistance in ovarian cancer positively correlates with enhanced Bcl-2 expression and reduced caspase-3 activity, but not with Bax and B-cell lymphoma-extra large (Bcl-xL) expression (Yang et al., 2002). In several cancer neoplasms, Bcl-2 is stabilized by FK506-binding proteins (FKBP38) and contributes to the development of chemoresistance (Fig. 3). The interaction between FKBP38 and flexible loop domain of Bcl-2 inhibits the phosphorylation of Bcl-2 and consequently prevents its degradation in chemoresistant cancer cells (Choi & Yoon, 2011). In leukemic cells, the IL-3-stimulated phosphorylation of Bcl-2 at serine$^{70}$ stabilizes the Bax–Bcl-2 interaction which suppresses drug-induced cell death (Deng, Kornblau, Ruvolo, & May, 2001) (Fig. 3). In cisplatin-resistant ovarian cancer cells, the Bcl-2 inhibitor ABT-737 (inhibits Bcl-2, Bcl-xL,

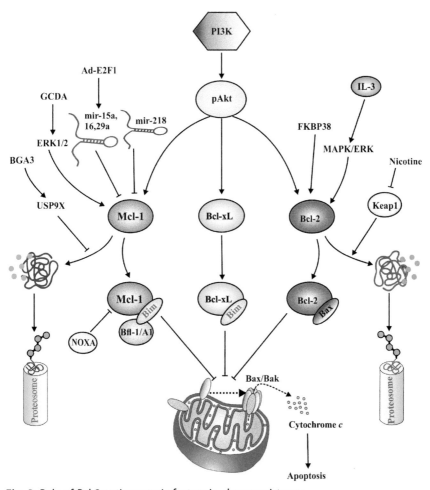

**Fig. 3** Role of Bcl-2 antiapoptotic factors in chemoresistance.

but not Mcl-1) sensitized these cancer cells to the antitumor efficacy of cisplatin (Dai, Jin, Li, & Wang, 2017). Nicotine, a major tobacco product, induces cisplatin resistance in lung cancer cells by stabilizing Bcl-2. Keap1 (an adaptor protein for Cul3-dependent protein) interacts with Bcl-2 and stimulates the ubiquitination and degradation of Bcl-2, but nicotine reduced this interaction resulting in stabilization of Bcl-2. In addition, nicotine also stabilizes Bcl-2 by activating Akt signaling (Nishioka et al., 2014). In multidrug-resistant lung cancer cells, Bcl-2 expression is upregulated, and knockdown of Bcl-2 using antisense RNA induces cisplatin-mediated apoptosis (Sartorius & Krammer, 2002). Estrogenic receptor (ER)-negative breast cancer cells were more sensitive to paclitaxel as compared to ER-positive cells. It was found that ER-regulated Bcl-2 expression was a key determinant of paclitaxel chemosensitivity in ER-positive breast cancer cells (Tabuchi et al., 2009). Knockdown of Bcl-2 by RNA interference decreases Bcl-2/Bax ratio and induces apoptosis in doxorubicin-resistant human osteosarcoma and chondrosarcoma cells (Kim, Kim, et al., 2009; Zhao et al., 2009).

## 2.3 Role of Mcl-1 in Chemoresistance

Overexpression or sustained Mcl-1 expression is a key determinant of cancer cell survival (Fig. 3). For decades it was believed that among all the antiapoptotic proteins Bcl-2 played an important role in drug resistance. Targeting Bcl-2 by ABT-737 (inhibits Bcl-2, Bcl-xL, Bcl-w) and ABT-199 showed remarkable efficacy against AML (Konopleva et al., 2006; Pan et al., 2014), but prolonged monotherapy led to acquired drug resistance against ABT-737/ABT-199. In ABT-737-sensitive cells, it displaced the proapoptotic Bim from Bcl-2 and induced cell death, but in acquired chemoresistant cells upregulation of Mcl-1/Bfl-1/A1 sequestered free Bim and inhibited apoptosis (Yecies, Carlson, Deng, & Letai, 2010) (Fig. 3). In ABT-199-resistant AML cells, inhibition of Mcl-1 and Bcl-xL results in restoration of ABT-199-induced cell death (Lin et al., 2016). In ABT-199-resistant cells, Mcl-1 and Bcl-xL are upregulated by AKT activation, which sequesters Bim. NVP-BEZ235, a dual inhibitor of AKT and mTOR, reduces Mcl-1 expression and sensitizes the ABT-199-resistant NHL cells to this therapy (Choudhary et al., 2015). Among all Bcl-2 antiapoptotic family members, only Mcl-1 was found to be upregulated in a majority of cisplatin-resistant cells. Inhibition of Mcl-1 genetically (siRNA) or pharmacologically (obatoclax) induces death in cisplatin-resistant cancer cells (Michels et al., 2014). Mcl-1 inhibition sensitizes chemoresistant neuroblastoma cells to etoposide, doxorubicin, and ABT-737 (Lestini et al., 2009).

EMT is one of the major events that occur during the process of acquired chemoresistance (Shintani et al., 2011). Inhibition of Mcl-1 in A549 cells using obatoclax resensitized EMT-induced chemoresistant cells to cisplatin toxicity (Toge et al., 2015). In carboplatin-resistant lung cancer cells, a tumor-suppressive role of miR-218 inversely correlated with Mcl-1 and survivin expression, which determines carboplatin sensitivity in lung cancer cells (Zarogoulidis et al., 2015) (Fig. 3). Telomerase-specific replication-competent oncolytic adenovirus (OBP-301) inhibits Mcl-1 and sensitized drug-resistant osteosarcoma cells to cisplatin and doxorubicin. OBP-301 induced miR-29 upregulation, which targets Mcl-1 by transcription factor E2F1 activation (Osaki et al., 2016). Enhanced expression of BAG3 (Bcl-2-associated athanogene 3) and Mcl-1 are key determinants for chemoresistance in ovarian cancer. The deubiquitinase activity of USP9X stabilizes Mcl-1 by inhibiting its proteasomal degradation (Schwickart et al., 2010) (Fig. 3). The interaction of BAG3 and USP9X stabilizes Mcl-1 and promotes resistance to drug-induced apoptosis. BAG3 knockdown reduces USP9X which subsequently suppresses Mcl-1 expression and enhances sensitivity to paclitaxel (Habata et al., 2016). A recent study showed that in hepatocellular carcinoma, GCDA (glycochenodeoxycholate) activated ERK1/ERK2, which phosphorylates Mcl-1 at T163 and increases the half-life of Mcl-1. Inhibition of Mcl-1 in HepG2 cells increases the sensitivity of hepatocellular cancer cells to cisplatin and irinotecan (Liao et al., 2011).

## 2.4 Role of Bcl-xL in Chemoresistance

Apart from Bcl-2 and Mcl-1, the antiapoptotic protein Bcl-xL is also reported to play an important role in developing drug resistance. In a cisplatin-resistant patient cohort of ovarian cancer, 61.5% of patients displayed enhanced Bcl-xL expression, 15% of patients had lower Bcl-xL expression, and 23% of patients had no change in Bcl-xL expression. From a nude mouse xenograft model, it was found that Bcl-xL-overexpressing tumors were resistant to cisplatin, paclitaxel, topotecan, and gemcitabine (Williams et al., 2005). Downregulation of Bcl-xL and Mcl-1 restored cell death even at low concentrations of cisplatin in recurrent and acquired chemoresistant ovarian cancer cells (Brotin et al., 2010). The Bcl-xL expression was determined in ovarian tumor tissues from 40 patient cohorts (20 taxane responsive and 20 with poor response to taxane). The majority of patients (10 out of 12) who were less responsive to taxane showed enhanced expression of Bcl-xL (Leibowitz & Yu, 2010). Similarly, cisplatin-resistant mesothelioma cells were sensitized to cisplatin upon knockdown of Bcl-xL (Varin et al., 2010).

## 2.5 Role of Bfl-1/A1 in Chemoresistance

Recently, detailed biochemical studies with recombinant Bcl-2 proteins and synthetic proapoptotic BH3 peptides revealed that previous assumptions about the relative affinities of hMcl-1 and hBfl-1 for NOXA may have not been entirely correct. In fact, it was demonstrated that hNOXA binds to human Bfl-1 using unique and conserved Cys residues, and, consequently, its affinity for hNOXA is over two orders of magnitude greater than that of hMcl-1, lacking a specific Cys residue in its binding site (Barile et al., 2017). These recent preliminary studies also indicate that a possible mechanism of regulation of hNOXA could require the formation of an intramolecular disulfide bridge between Cys residues located in its BH3 region and at the base of the TM domain, necessary for localization in the mitochondrial membrane. hNOXA is the only proapoptotic factor among the Bcl-2 proteins that is activated by UV radiation (Naik, Michalak, Villunger, Adams, & Strasser, 2007), perhaps because UV may catalyze the disulfide bridge opening, hence exposing the TM domain allowing it to translocate to the mitochondria. Furthermore, dysfunctional mitochondria in cancer cells can increase the generation of ROS (reactive oxygen species), and such altered redox potential in cancer cells could be a target for both hNOXA activation and interference with hNOXA/hBfl-1 interactions (Barile et al., 2017). For example, Bfl-1 is often found overexpressed in melanoma cells lines (Placzek et al., 2010), and in most solid tumors, the expression rate of Bfl-1 in metastatic lymph nodes is 82%, which is higher than 50% in the primary sites ($P < 0.02$) (Park et al., 1997). Moreover, in chronic lymphocytic leukemia (CLL), recent studies suggest that Bfl-1 may have a more dominant role in resistance to both chemotherapy and Bcl-2 antagonists than hMcl-1 (Olsson et al., 2007; Yecies et al., 2010). These observations, together with the recent findings that hNOXA possesses intrinsically a greater affinity for hBfl-1 compared to hMcl-1, make hBfl-1 potentially a very intriguing potential target for pharmaceutical intervention (Barile et al., 2017; Morales et al., 2005; Vogler, 2012; Vogler et al., 2009).

## 3. POTENTIAL Bcl-2 INHIBITORS IN CHEMORESISTANCE: CLINICAL DEVELOPMENT

Several clinical trials were conducted using Bcl-2 inhibitors, where these molecules were used as either a single agent or in combination with chemotherapeutic drugs in case of various relapsed or refractory neoplasms (Table 1). A detailed list of completed and on-going trials is listed in Table 1.

### 3.1 ABT-263 (Navitoclax)

ABT-263 is an orally deliverable potent inhibitor of Bcl-xL, Bcl-2, and Bcl-w, which binds weakly with Mcl-1. A phase II study was conducted with the combination of ABT-263/abiraterone or ABT-263/abiraterone/hydroxychloroquine where patients with metastatic castrate refractory prostate cancer were enrolled. The study has been terminated, but no outcome has been published (NCT01828476). In a phase II clinical trial, the antitumor efficacy of ABT-263 in combination with bendamustine and rituximab was evaluated in patients with relapsed diffuse large B-cell lymphoma, but the study was withdrawn prior to its evaluation (NCT01423539).

### 3.2 ABT-199 (Venetoclax)

Venetoclax is an orally bioavailable BH3 mimetic that specifically inhibits Bcl-2 protein. A phase I clinical trial was conducted with a single dose of ABT-199 in patients with refractory NHL, but the study was withdrawn prior to enrollment (NCT02095574). A phase I clinical trial was conducted using ABT-199 in combination with ibrutinib in relapsed mantle cell lymphoma (MCL) patients, where the dose-limiting toxicities were determined (NCT02419560).

### 3.3 AT-101 (Gossypol)

AT-101 is an orally administered inhibitor of Bcl-2 antiapoptotic proteins (Bcl-2, Bcl-xL, Bcl-W, and Mcl-1). Using AT-101 as a single agent, a phase II trial was conducted where 29 subjects were enrolled for recurrent adrenocortical cancer that could not be removed by surgery. In this study, AT-101 showed antitumor efficacy by blocking the growth of tumors (NCT00848016). In a phase II trial, AT-101 stopped the growth of tumor cells by blocking some of the essential enzymes needed for cell growth in 56 recurrent glioblastoma patients (NCT00540722). Similarly, in a phase

Table 1 Ongoing and Completed Clinical Trials With Bcl-2 Inhibitors in Refractory Neoplasms

| Bcl-2 Inhibitors | Combination With | Relapsed/Refractory | Phase | Status | Clinical Trials.gov Identifier | References |
|---|---|---|---|---|---|---|
| ABT-263 (navitoclax) | As a single agent | Ovarian cancer | Phase II | Active, not recruiting | NCT02591095 | |
| | Fludarabine/cyclophosphamide or rituximab (FCR) or bendamustine/rituximab (BR) | CLL | Phase I | Completed | NCT00868413 | |
| | As a single agent | Lymphoid malignancies | Phase II | Completed | NCT00406809 | PMID: 21094089 |
| | As a single agent | CLL | Phase II | Active, not recruiting | NCT00481091 | |
| | Sorafenib tosylate | Hepatocellular carcinoma | Phase I | Suspended | NCT02143401 | |
| | Trametinib | Colorectal carcinoma, lung carcinoma, pancreatic carcinoma | Phase I/II | Recruiting | NCT02079740 | |
| | Abiraterone, hydroxychloroquine | Prostate cancer | Phase II | Terminated | NCT01822476 | |
| | Osimertinib | NSCLC | Phase I | Recruiting | NCT02520778 | |
| | Dabrafenib, trametinib | Melanoma | Phase I/II | Recruiting | NCT01989585 | |
| | Bendamustine, rituximab | Large B-cell lymphoma | Phase II | Withdrawn | NCT01423539 | |

| ABT-199 (venetoclax) | Ibrutinib | Lymphoma, mantle cell | Phase I | Suspended | NCT02419560 | |
| --- | --- | --- | --- | --- | --- | --- |
| | Single agent | NHL | Phase I | Withdrawn | NCT02095574 | |
| | Single agent | Waldenstrom's macroglobulinemia | Phase II | Recruiting | NCT02677324 | |
| | Rituximab | CML | Phase I | Active, not recruiting | NCT01682616 | PMID: 28089635 |
| | As a single agent | CLL | Phase II | Active, not recruiting | NCT02141282 | |
| | As a single agent | CLL and NHL | Phase I | Active, not recruiting | NCT01328626 | PMID: 26639348 |
| | Bortezomib, dexamethasone | MM | Phase I | Active, not recruiting | NCT01794507 | |
| | Dexamethasone | MM | Phase I | Active, not recruiting | NCT01794520 | |
| | Bendamustine, rituximab | NHL | Phase I | Active, not recruiting | NCT01594229 | |
| | As a single agent | CLL | Phase II | Active, not recruiting | NCT01889186 | PMID: 27178240 |
| | Ibrutinib | MCL | Phase I/II | Recruiting | NCT02255816 | |
| | Bendamustine, rituximab | CLL | Phase III | Active, not recruiting | NCT02005471 | |
| | Bendamustine, GDC-0199, rituximab | fNHL | Phase II | Active, not recruiting | NCT02187861 | |
| | As a single agent | CLL | Phase III | Recruiting | NCT02756611 | |
| | Bendamustine, obinutuzumab, rituximab | CLL | Phase I | Recruiting | NCT01671904 | |

*Continued*

**Table 1** Ongoing and Completed Clinical Trials With Bcl-2 Inhibitors in Refractory Neoplasms—cont'd

| Bcl-2 Inhibitors | Combination With | Relapsed/Refractory | Phase | Status | Clinical Trials.gov Identifier | References |
|---|---|---|---|---|---|---|
| | Obinutuzumab, ibrutinib | CLL | Phase I/II | Active, not recruiting | NCT02427451 | |
| | Obinutuzumab | DLBCL | Phase II | Recruiting | NCT02987400 | |
| | As a single agent | CLL | Phase II | Not yet recruiting | NCT02966756 | |
| | Dexamethasone, carfilzomib | MM | Phase II | Recruiting | NCT02899052 | |
| | Ibrutinib | Follicular lymphoma | Phase I/II | Recruiting | NCT02956382 | |
| | Ibrutinib | CLL or SLL | Phase I/II | Not yet recruiting | NCT03045328 | |
| | Lenalidomide, obinutuzumab | B-cell NHL | Phase I | Recruiting | NCT02992522 | |
| AT-101 | Single agent | Adrenocortical carcinoma | Phase II | Completed | NCT00848016 | |
| | Lenalidomide | B-cell CLL | Phase I/II | Recruiting | NCT01003769 | |
| | Docetaxel | NSCLC | Phase II | Completed | NCT00544960 | |
| | Prednisone and docetaxel | H-Prostate cancer | Phase II | Completed | NCT00571675 | PMID: 21982118 |
| | Topotecan | SCLC | Phase I/II | Completed | NCT00397293 | |
| | As a single agent | B-cell malignancies | Phase II | Completed | NCT00275431 | |
| | As a single agent | Glioblastoma multiforme | Phase II | Completed | NCT00540722 | |
| | As a single agent | H-Prostate cancer | Phase I/II | Completed | NCT00286806 | |
| | As a single agent | Extensive-stage SCLC | Phase II | Completed | NCT00773955 | |
| | Dexamethasone, lenalidomide | Symptomatic MM | Phase I/II | Recruiting | NCT02697344 | |

| | | | | | |
|---|---|---|---|---|---|
| Flavopiridol | As a single agent | MCL | Phase I/II | Completed | NCT00445341 | PMID: 15930354 |
| | Alvocidib | AML | Phase I | Terminated | NCT00101231 | PMID: 20460644 |
| | Alvocidib | CLL or SLL | Phase I/II | Completed | NCT00058240 | |
| | Alvocidib | Lymphoma or MM | Phase I/II | Terminated | NCT00112723 | PMID: 24241210 |
| | Alvocidib | CLL | Phase II | Completed | NCT00003620 | |
| | Alvocidib | CLL | Phase II | Terminated | NCT00098371 | PMID: 22289993 |
| | Alvocidib | MM | Phase II | Completed | NCT00047203 | PMID: 25596730 |
| | Alvocidib, vorinostat | Acute leukemia or CML | Phase I | Completed | NCT00278330 | |
| | Alvocidib | Solid tumors or lymphomas | Phase I | Completed | NCT00012181 | |
| | Alvocidib, docetaxel | Metastatic pancreatic cancer | Phase II | Completed | NCT00331682 | |
| | Alvocidib, cisplatin | Recurrent ovarian cancer | Phase II | Completed | NCT00083122 | |
| | Alvocidib, fluorouracil, leucovorin calcium, oxaliplatin | Metastatic or recurrent sarcoma | Phase I | Completed | NCT00098579 | |
| | Paclitaxel | Metastatic esophageal cancer | Phase II | Completed | NCT00006245 | |
| | Acetylsalicylic acid, alvocidib, clopidogrel bisulfate | Recurrent or metastatic HNC | Phase II | Completed | NCT00020189 | |
| | Alvocidib, mitoxantrone hydrochloride, carboplatin, | AML | Phase II | Recruiting | NCT00634244 | |

*Continued*

Table 1 Ongoing and Completed Clinical Trials With Bcl-2 Inhibitors in Refractory Neoplasms—cont'd

| Bcl-2 Inhibitors | Combination With | Relapsed/Refractory | Phase | Status | Clinical Trials.gov Identifier | References |
|---|---|---|---|---|---|---|
| | cytarabine, sirolimus, etoposide, topotecan hydrochloride | | | | | |
| | Cisplatin, paclitaxel | Solid tumors | Phase I | Completed | NCT00003004 | |
| Gossypol | As a single agent | Recurrent, metastatic, or primary adrenocortical cancer | Phase II | Completed | NCT00848016 | |
| | As a single agent | Progressive or recurrent glioblastoma multiforme | Phase II | Completed | NCT00540722 | |
| | As a single agent | Extensive-stage SCLC | Phase II | Completed | NCT00773955 | |
| | Lenalidomide | B-cell CLL | Phase I/II | Recruiting | NCT01003769 | |
| | Dexamethasone, lenalidomide | Symptomatic MM | Phase I/II | Recruiting | NCT02697344 | |
| Obatoclax | Bortezomib | NHL | Phase I | Terminated | NCT00538187 | |
| | Bortezomib | MCL | Phase I/II | Completed | NCT00407303 | |
| | Dexrazoxane hydrochloride, doxorubicin hydrochloride, liposomal vincristine sulfate | Solid tumors, lymphoma, or leukemia | Phase I | Terminated | NCT00933985 | |
| | Bortezomib | MM | Phase I/II | Terminated | NCT00719901 | |
| | Topotecan hydrochloride | SCLC or advanced solid tumors | Phase I/II | Completed | NCT00521144 | PMID: 21620511 |
| | Bendamustine hydrochloride, Rituximab | NHL | Phase I/II | Withdrawn | NCT01238146 | |

| | | | | | |
|---|---|---|---|---|---|
| Oblimersen (G-3139) | Single agent | Waldenstrom's macroglobulinemia | Phase I/II | Completed | NCT00062244 |
| | Single agent | Neuroendocrine carcinoma of the skin | Phase II | Completed | NCT00079131 |
| | Gemcitabine hydrochloride | Lymphoma | Phase I | Terminated | NCT00060112 |
| | Rituximab | B-cell NHL | Phase II | Completed | NCT00054639 |
| | Interferon alfa | Renal cell cancer | Phase II | Completed | NCT00059813 |
| | Docetaxel | H-adenocarcinoma (cancer) of the prostate | Phase II | Completed | NCT00085228 PMID: 19297314 |
| | Dexrazoxane, doxorubicin, cyclophosphamide, filgrastim | Solid tumors | Phase I | Completed | NCT00039481 |
| | Gemtuzumab ozogamicin | AML | Phase II | Completed | NCT00017589 PMID: 16730060 |
| | Dexamethasone | MM | Phase III | Completed | NCT00017602 PMID: 19373653 |
| | Paclitaxel | SCLC | Phase I/II | Completed | NCT00005032 PMID: 12056703 |
| | Dexamethasone, thalidomide | MM | Phase II | Completed | NCT00049374 PMID: 15867202 |
| | Filgrastim, cyclophosphamide, fludarabine phosphate | CLL | Phase III | Completed | NCT00024440 PMID: 19738118 PMID: 17296974 |
| | Rituximab, ifosfamide, carboplatin, etoposide, filgrastim, pegfilgrastim | Aggressive NHL | Phase I/II | Completed | NCT00086944 |

II clinical trial, the enantiomer of AT-101 (R-(−)-gossypol acetic acid) was used in case of extensive-stage small-cell lung cancer (ECMC) where it stopped the growth of tumor cells (NCT00773955). In a phase II study, 106 subjects were enrolled to evaluate the efficacy of AT-101 in combination with docetaxel in relapsed NSCLC. The aim of the study was to estimate and compare the progression-free survival of AT-101 in combination with docetaxel and placebo (NCT00544960). In a multinational phase II clinical trial, AT-101 was administered orally in combination with docetaxel and prednisone to hormone-refractory prostate cancer patients (NCT00571675). Topotecan in combination with AT-101 was evaluated in a multicenter phase I/II clinical trial in patients with refractory SCLC (NCT00397293).

### 3.4 GX15-070 (Obatoclax)

GX15-070 is a small hydrophobic indole bipyrrole compound that antagonizes Bcl-2, Bcl-xL, Bcl-w, and Mcl-1. In a phase I/II clinical trial 18 patients were recruited with aggressive recurrent NHL, and the combinatorial efficacy of bortezomib and obatoclax was evaluated. In this study, the maximum tolerated dose (MTD), toxicity, pharmacokinetic behavior, and clinical responses were determined (NCT00538187). In another phase I/II study, obatoclax was administered in combination with bortezomib to patients with refractory MCL. Twenty-four patients were enrolled for this trial where the combination treatment was designed to restore apoptosis through inhibition of the Bcl-2 family of proteins (NCT00407303). In a phase I/II trial, effects and best dose of obatoclax together with bortezomib were evaluated in 11 patients with refractory multiple myeloma. The response rate (complete response, partial response, and very good partial response) in patients treated with this regimen was determined along with the duration of progression-free, overall survival and incidence of toxicities (NCT00719901).

### 3.5 Oblimersen

Oblimersen sodium is a Bcl-2 antisense oligonucleotide which increases myeloma cell susceptibility to cytotoxic agents. In a phase I/II clinical trial, oblimersen was administered as a single agent in 58 relapsed Waldenstrom's macroglobulinemia patients. In this study, oblimersen blocked the growth of the cancer cells (NCT00062244). In a phase I trial, the combinatorial effect of oblimersen and gemcitabine was evaluated in 15 patients with recurrent lymphoma, where the MTD was successfully evaluated (NCT00060112).

Similarly, the toxicity profile of the combination of oblimersen and rituximab was evaluated in patients with recurrent B-cell NHL (NCT00054639). In another randomized phase II trial, the toxicity profile of docetaxel together with oblimersen was evaluated in patients with hormone-refractory prostate cancer. PSA response in patients showed 46% and 36% with 57 and 54 patients treated with docetaxel and docetaxel–oblimersen, respectively (NCT00085228). Also in a randomized phase III trial, the effectiveness of dexamethasone with or without oblimersen was evaluated in patients with relapsed multiple myeloma. In this study, 110 patients received the combination treatment and 114 patients received dexamethasone alone. The results demonstrated no significant differences between the two groups relative to time to tumor progression (NCT00017602). In another phase II trial, the effectiveness of thalidomide and dexamethasone with oblimersen was evaluated in patients with relapsed multiple myeloma. The study results suggested that thalidomide, dexamethasone, and thalidomide are well tolerated and result in encouraging clinical responses in relapsed patients (NCT00049374). In another randomized phase III clinical trial, fludarabine and cyclophosphamide with or without oblimersen were administered to patients who have relapsed CLL. The study result suggested that partial remission could be achieved with this combination in relapsed CLL (NCT00024440).

## 4. CONCLUSIONS AND FUTURE PERSPECTIVES

As discussed in this chapter, several preclinical and clinical studies have been conducted to understand the role of Bcl-2 family proteins and reduced apoptotic response in chemoresistant cancer cells. In principle, Bcl-2 antiapoptotic proteins are potential therapeutic targets to restore cell death in chemoresistant cancers. Bcl-2 inhibitors either alone or in combination with existing chemotherapeutics can overcome therapy-resistant/recurrent tumors to increase the disease-free survival of cancer patients. Further studies are required to confirm the true clinical potential of Bcl-2 inhibitors as single and combinatorial agents for the therapy of chemotherapy-sensitive and resistant cancers.

## ACKNOWLEDGMENTS

This work is supported by the Department of Biotechnology, DST (EMR/2015/000063), India (R.D.), the National Institutes of Health, NCI grant R01 CA168517 (M.P. and P.B.F.), and the National Foundation for Cancer Research (NFCR) (P.B.F.). R.D. is thankful for the Ramalingaswami Fellowship. M.P. holds the Daniel Hays Chair in Cancer Research at the School of Medicine at UCR. P.B.F. holds the Thelma Newmeyer Corman Chair in Cancer Research at the VCU Massey Cancer Center of VCU School of Medicine.

## REFERENCES

Abdullah, L. N., & Chow, E. K. (2013). Mechanisms of chemoresistance in cancer stem cells. *Clinical and Translational Medicine, 2,* 3.

Adhikari, S., Choudhury, S., Mitra, P. S., Dubash, J. J., Sajankila, S. P., & Roy, R. (2008). Targeting base excision repair for chemosensitization. *Anti-Cancer Agents in Medicinal Chemistry, 8,* 351–357.

Ashkenazi, A., & Dixit, V. M. (1999). Apoptosis control by death and decoy receptors. *Current Opinion in Cell Biology, 11,* 255–260.

Azuma, Y., Yokobori, T., Mogi, A., Altan, B., Yajima, T., Kosaka, T., et al. (2015). SIRT6 expression is associated with poor prognosis and chemosensitivity in patients with non-small cell lung cancer. *Journal of Surgical Oncology, 112,* 231–237.

Bae, J., Leo, C. P., Hsu, S. Y., & Hsueh, A. J. (2000). MCL-1S, a splicing variant of the antiapoptotic BCL-2 family member MCL-1, encodes a proapoptotic protein possessing only the BH3 domain. *The Journal of Biological Chemistry, 275,* 25255–25261.

Bakhshi, A., Jensen, J. P., Goldman, P., Wright, J. J., McBride, O. W., Epstein, A. L., et al. (1985). Cloning the chromosomal breakpoint of t(14;18) human lymphomas: Clustering around JH on chromosome 14 and near a transcriptional unit on 18. *Cell, 41,* 899–906.

Balaji, S. A., Udupa, N., Chamallamudi, M. R., Gupta, V., & Rangarajan, A. (2016). Role of the drug transporter ABCC3 in breast cancer chemoresistance. *PLoS One, 11,* e0155013.

Barile, E., Marconi, G. D., De, S. K., Baggio, C., Gambini, L., Salem, A. F., et al. (2017). hBfl-1/hNOXA interaction studies provide new insights on the role of Bfl-1 in cancer cell resistance and for the design of novel anticancer agents. *ACS Chemical Biology, 12,* 444–455.

Bartek, J., Bartkova, J., & Lukas, J. (2007). DNA damage signalling guards against activated oncogenes and tumour progression. *Oncogene, 26,* 7773–7779.

Berger, S. L., Kouzarides, T., Shiekhattar, R., & Shilatifard, A. (2009). An operational definition of epigenetics. *Genes & Development, 23,* 781–783.

Bhutia, S. K., Mukhopadhyay, S., Sinha, N., Das, D. N., Panda, P. K., Patra, S. K., et al. (2013). Autophagy: Cancer's friend or foe? *Advances in Cancer Research, 118,* 61–95.

Blagosklonny, M. V. (2006). Target for cancer therapy: Proliferating cells or stem cells. *Leukemia, 20,* 385–391.

Bockhorn, J., Dalton, R., Nwachukwu, C., Huang, S., Prat, A., Yee, K., et al. (2013). MicroRNA-30c inhibits human breast tumour chemotherapy resistance by regulating TWF1 and IL-11. *Nature Communications, 4,* 1393.

Bose, D., Zimmerman, L. J., Pierobon, M., Petricoin, E., Tozzi, F., Parikh, A., et al. (2011). Chemoresistant colorectal cancer cells and cancer stem cells mediate growth and survival of bystander cells. *British Journal of Cancer, 105,* 1759–1767.

Brotin, E., Meryet-Figuiere, M., Simonin, K., Duval, R. E., Villedieu, M., Leroy-Dudal, J., et al. (2010). Bcl-XL and MCL-1 constitute pertinent targets in ovarian carcinoma and their concomitant inhibition is sufficient to induce apoptosis. *International Journal of Cancer, 126,* 885–895.

Cacan, E. (2017). Epigenetic-mediated immune suppression of positive co-stimulatory molecules in chemoresistant ovarian cancer cells. *Cell Biology International, 41,* 328–339.

Cai, B. L., Xu, X. F., Fu, S. M., Shen, L. L., Zhang, J., Guan, S. M., et al. (2011). Nuclear translocation of MRP1 contributes to multidrug resistance of mucoepidermoid carcinoma. *Oral Oncology, 47,* 1134–1140.

Calabrese, C. R., Almassy, R., Barton, S., Batey, M. A., Calvert, A. H., Canan-Koch, S., et al. (2004). Anticancer chemosensitization and radiosensitization by the novel poly(ADP-ribose) polymerase-1 inhibitor AG14361. *Journal of the National Cancer Institute*, *96*, 56–67.

Campbell, K. J., Bath, M. L., Turner, M. L., Vandenberg, C. J., Bouillet, P., Metcalf, D., et al. (2010). Elevated Mcl-1 perturbs lymphopoiesis, promotes transformation of hematopoietic stem/progenitor cells, and enhances drug resistance. *Blood*, *116*, 3197–3207.

Campos, L., Rouault, J. P., Sabido, O., Oriol, P., Roubi, N., Vasselon, C., et al. (1993). High expression of bcl-2 protein in acute myeloid leukemia cells is associated with poor response to chemotherapy. *Blood*, *81*, 3091–3096.

Cao, X., Fang, L., Gibbs, S., Huang, Y., Dai, Z., Wen, P., et al. (2007). Glucose uptake inhibitor sensitizes cancer cells to daunorubicin and overcomes drug resistance in hypoxia. *Cancer Chemotherapy and Pharmacology*, *59*, 495–505.

Chabner, B. A., & Roberts, T. G., Jr. (2005). Timeline: Chemotherapy and the war on cancer. *Nature Reviews Cancer*, *5*, 65–72.

Chen, D., Wu, M., Li, Y., Chang, I., Yuan, Q., Ekimyan-Salvo, M., et al. (2017). Targeting BMI1+ cancer stem cells overcomes chemoresistance and inhibits metastases in squamous cell carcinoma. *Cell Stem Cell*, *20*(621–634), e6.

Chen, W. T., Yang, Y. J., Zhang, Z. D., An, Q., Li, N., Liu, W., et al. (2017). MiR-1307 promotes ovarian cancer cell chemoresistance by targeting the ING5 expression. *Journal of Ovarian Research*, *10*(1).

Chen, H., Zhang, W., Cheng, X., Guo, L., Xie, S., Ma, Y., et al. (2017). beta2-AR activation induces chemoresistance by modulating p53 acetylation through upregulating Sirt1 in cervical cancer cells. *Cancer Science*, *108*, 1310–1317.

Chen, B., Zhang, D., Kuai, J., Cheng, M., Fang, X., & Li, G. (2017). Upregulation of miR-199a/b contributes to cisplatin resistance via Wnt/beta-catenin-ABCG2 signaling pathway in ALDHA1+ colorectal cancer stem cells. *Tumour Biology*, *39*, 1010428317715155.

Chiang, Y. C., Chang, M. C., Chen, P. J., Wu, M. M., Hsieh, C. Y., Cheng, W. F., et al. (2013). Epigenetic silencing of BLU through interfering apoptosis results in chemoresistance and poor prognosis of ovarian serous carcinoma patients. *Endocrine-Related Cancer*, *20*, 213–227.

Choi, B. Y., Joo, J. C., Lee, Y. K., Jang, I. S., Park, S. J., & Park, Y. J. (2017). Anti-cancer effect of Scutellaria baicalensis in combination with cisplatin in human ovarian cancer cell. *BMC Complementary and Alternative Medicine*, *17*, 277.

Choi, B. H., & Yoon, H. S. (2011). FKBP38-Bcl-2 interaction: A novel link to chemoresistance. *Current Opinion in Pharmacology*, *11*, 354–359.

Choudhary, G. S., Al-Harbi, S., Mazumder, S., Hill, B. T., Smith, M. R., Bodo, J., et al. (2015). MCL-1 and BCL-xL-dependent resistance to the BCL-2 inhibitor ABT-199 can be overcome by preventing PI3K/AKT/mTOR activation in lymphoid malignancies. *Cell Death & Disease*, *6*, e1593.

Coloff, J. L., Macintyre, A. N., Nichols, A. G., Liu, T., Gallo, C. A., Plas, D. R., et al. (2011). Akt-dependent glucose metabolism promotes Mcl-1 synthesis to maintain cell survival and resistance to Bcl-2 inhibition. *Cancer Research*, *71*, 5204–5213.

Crea, F., Nobili, S., Paolicchi, E., Perrone, G., Napoli, C., Landini, I., et al. (2011). Epigenetics and chemoresistance in colorectal cancer: An opportunity for treatment tailoring and novel therapeutic strategies. *Drug Resistance Updates*, *14*, 280–296.

Croce, C. M., & Fisher, P. B. (2017). MicroRNA and cancer. *Advances in Cancer Research*, *135*, ix–xi.

Dai, Y., Jin, S., Li, X., & Wang, D. (2017). The involvement of Bcl-2 family proteins in AKT-regulated cell survival in cisplatin resistant epithelial ovarian cancer. *Oncotarget*, *8*, 1354–1368.

Dalvi, M. P., Wang, L., Zhong, R., Kollipara, R. K., Park, H., Bayo, J., et al. (2017). Taxane-platin-resistant lung cancers co-develop hypersensitivity to JumonjiC demethylase inhibitors. *Cell Reports, 19*, 1669–1684.

Danial, N. N. (2007). BCL-2 family proteins: Critical checkpoints of apoptotic cell death. *Clinical Cancer Research, 13*, 7254–7263.

Dash, R., Azab, B., Quinn, B. A., Shen, X., Wang, X. Y., Das, S. K., et al. (2011). Apogossypol derivative BI-97C1 (Sabutoclax) targeting Mcl-1 sensitizes prostate cancer cells to mda-7/IL-24-mediated toxicity. *Proceedings of the National Academy of Sciences of the United States of America, 108*, 8785–8790.

Dash, R., Richards, J. E., Su, Z. Z., Bhutia, S. K., Azab, B., Rahmani, M., et al. (2010). Mechanism by which Mcl-1 regulates cancer-specific apoptosis triggered by mda-7/IL-24, an IL-10-related cytokine. *Cancer Research, 70*, 5034–5045.

Dean, M. (2009). ABC transporters, drug resistance, and cancer stem cells. *Journal of Mammary Gland Biology and Neoplasia, 14*, 3–9.

de la Cruz-Morcillo, M. A., Valero, M. L., Callejas-Valera, J. L., Arias-Gonzalez, L., Melgar-Rojas, P., Galan-Moya, E. M., et al. (2012). P38MAPK is a major determinant of the balance between apoptosis and autophagy triggered by 5-fluorouracil: Implication in resistance. *Oncogene, 31*, 1073–1085.

Deng, X., Kornblau, S. M., Ruvolo, P. P., & May, W. S., Jr. (2001). Regulation of Bcl2 phosphorylation and potential significance for leukemic cell chemoresistance. *Journal of the National Cancer Institute Monographs*, 30–37.

Deng, J., Wang, L., Chen, H., Hao, J., Ni, J., Chang, L., et al. (2016). Targeting epithelial-mesenchymal transition and cancer stem cells for chemoresistant ovarian cancer. *Oncotarget, 7*, 55771–55788.

de Souza Viana, L., de Aguiar Silva, F. C., Andrade Dos Anjos Jacome, A., Calheiros Campelo Maia, D., Duarte de Mattos, M., Arthur Jacinto, A., et al. (2016). Efficacy and safety of a cisplatin and paclitaxel induction regimen followed by chemoradiotherapy for patients with locally advanced head and neck squamous cell carcinoma. *Head & Neck, 38*(Suppl. 1), E970–80.

Doles, J., Oliver, T. G., Cameron, E. R., Hsu, G., Jacks, T., Walker, G. C., et al. (2010). Suppression of Rev3, the catalytic subunit of Pol{zeta}, sensitizes drug-resistant lung tumors to chemotherapy. *Proceedings of the National Academy of Sciences of the United States of America, 107*, 20786–20791.

Donawho, C. K., Luo, Y., Luo, Y., Penning, T. D., Bauch, J. L., Bouska, J. J., et al. (2007). ABT-888, an orally active poly(ADP-ribose) polymerase inhibitor that potentiates DNA-damaging agents in preclinical tumor models. *Clinical Cancer Research, 13*, 2728–2737.

du Bois, A., Luck, H. J., Meier, W., Adams, H. P., Mobus, V., Costa, S., et al. (2003). A randomized clinical trial of cisplatin/paclitaxel versus carboplatin/paclitaxel as first-line treatment of ovarian cancer. *Journal of the National Cancer Institute, 95*, 1320–1329.

Elkholi, R., Floros, K. V., & Chipuk, J. E. (2011). The role of BH3-only proteins in tumor cell development, signaling, and treatment. *Genes Cancer, 2*, 523–537.

Evers, B., Drost, R., Schut, E., de Bruin, M., van der Burg, E., Derksen, P. W., et al. (2008). Selective inhibition of BRCA2-deficient mammary tumor cell growth by AZD2281 and cisplatin. *Clinical Cancer Research, 14*, 3916–3925.

Filipowicz, W., Bhattacharyya, S. N., & Sonenberg, N. (2008). Mechanisms of post-transcriptional regulation by microRNAs: Are the answers in sight? *Nature Reviews Genetics, 9*, 102–114.

Flavin, R., Peluso, S., Nguyen, P. L., & Loda, M. (2010). Fatty acid synthase as a potential therapeutic target in cancer. *Future Oncology, 6*, 551–562.

Frei, E., 3rd, Karon, M., Levin, R. H., Freireich, E. J., Taylor, R. J., Hananian, J., et al. (1965). The effectiveness of combinations of antileukemic agents in inducing and maintaining remission in children with acute leukemia. *Blood, 26*, 642–656.

Fujita, Y., Yagishita, S., Hagiwara, K., Yoshioka, Y., Kosaka, N., Takeshita, F., et al. (2015). The clinical relevance of the miR-197/CKS1B/STAT3-mediated PD-L1 network in chemoresistant non-small-cell lung cancer. *Molecular Therapy, 23*, 717–727.

Galluzzi, L., Bravo-San Pedro, J. M., Levine, B., Green, D. R., & Kroemer, G. (2017). Pharmacological modulation of autophagy: Therapeutic potential and persisting obstacles. *Nature Reviews Drug Discovery, 16*, 487–511.

Gao, W., Li, J. Z., Chen, S. Q., Chu, C. Y., Chan, J. Y., & Wong, T. S. (2017). BEX3 contributes to cisplatin chemoresistance in nasopharyngeal carcinoma. *Cancer Medicine, 6*, 439–451.

Gao, C., Zhang, J., Wang, Q., & Ren, C. (2016). Overexpression of lncRNA NEAT1 mitigates multidrug resistance by inhibiting ABCG2 in leukemia. *Oncology Letters, 12*, 1051–1057.

Garofalo, M., Romano, G., Di Leva, G., Nuovo, G., Jeon, Y. J., Ngankeu, A., et al. (2011). EGFR and MET receptor tyrosine kinase-altered microRNA expression induces tumorigenesis and gefitinib resistance in lung cancers. *Nature Medicine, 18*, 74–82.

Garza-Trevino, E. N., Said-Fernandez, S. L., & Martinez-Rodriguez, H. G. (2015). Understanding the colon cancer stem cells and perspectives on treatment. *Cancer Cell International, 15*, 2.

Geng, M., Wang, L., & Li, P. (2013). Correlation between chemosensitivity to anticancer drugs and Bcl-2 expression in gastric cancer. *International Journal of Clinical and Experimental Pathology, 6*, 2554–2559.

Ghuwalewala, S., Ghatak, D., Das, P., Dey, S., Sarkar, S., Alam, N., et al. (2016). CD44 (high)CD24(low) molecular signature determines the cancer stem cell and EMT phenotype in oral squamous cell carcinoma. *Stem Cell Research, 16*, 405–417.

Gilman, A. (1963). The initial clinical trial of nitrogen mustard. *American Journal of Surgery, 105*, 574–578.

Go, R. S., & Adjei, A. A. (1999). Review of the comparative pharmacology and clinical activity of cisplatin and carboplatin. *Journal of Clinical Oncology, 17*, 409–422.

Goldberg, S. B., Supko, J. G., Neal, J. W., Muzikansky, A., Digumarthy, S., Fidias, P., et al. (2012). A phase I study of erlotinib and hydroxychloroquine in advanced non-small-cell lung cancer. *Journal of Thoracic Oncology, 7*, 1602–1608.

Gottesman, M. M. (2002). Mechanisms of cancer drug resistance. *Annual Review of Medicine, 53*, 615–627.

Green, D. R., & Kroemer, G. (2004). The pathophysiology of mitochondrial cell death. *Science, 305*, 626–629.

Grover, R., & Wilson, G. D. (1996). Bcl-2 expression in malignant melanoma and its prognostic significance. *European Journal of Surgical Oncology, 22*, 347–349.

Gu, X., Lu, Y., He, D., Lu, C., Jin, J., Lu, X., et al. (2016). Methylation of Notch3 modulates chemoresistance via P-glycoprotein. *European Journal of Pharmacology, 792*, 7–14.

Guo, W., Lasky, J. L., 3rd, & Wu, H. (2006). Cancer stem cells. *Pediatric Research, 59*, 59R–64R.

Habata, S., Iwasaki, M., Sugio, A., Suzuki, M., Tamate, M., Satohisa, S., et al. (2016). BAG3-mediated Mcl-1 stabilization contributes to drug resistance via interaction with USP9X in ovarian cancer. *International Journal of Oncology, 49*, 402–410.

Ham, S. Y., Kwon, T., Bak, Y., Yu, J. H., Hong, J., Lee, S. K., et al. (2016). Mucin 1-mediated chemo-resistance in lung cancer cells. *Oncogenesis, 5*, e185.

Han, W., Pan, H., Chen, Y., Sun, J., Wang, Y., Li, J., et al. (2011). EGFR tyrosine kinase inhibitors activate autophagy as a cytoprotective response in human lung cancer cells. *PLoS One, 6*, e18691.

He, X., Wang, J., Wei, W., Shi, M., Xin, B., Zhang, T., et al. (2016). Hypoxia regulates ABCG2 activity through the activation of ERK1/2/HIF-1alpha and contributes to chemoresistance in pancreatic cancer cells. *Cancer Biology & Therapy, 17*, 188–198.

Hermine, O., Haioun, C., Lepage, E., d'Agay, M. F., Briere, J., Lavignac, C., et al. (1996). Prognostic significance of bcl-2 protein expression in aggressive non-Hodgkin's lymphoma. Groupe d'Etude des Lymphomes de l'Adulte (GELA). *Blood, 87*, 265–272.

Hitchings, G. H., & Elion, G. B. (1954). The chemistry and biochemistry of purine analogs. *Annals of the New York Academy of Sciences, 60*, 195–199.

Hsu, H. S., Lin, J. H., Huang, W. C., Hsu, T. W., Su, K., Chiou, S. H., et al. (2011). Chemoresistance of lung cancer stemlike cells depends on activation of Hsp27. *Cancer, 117*, 1516–1528.

Hu, Y., Yan, C., Mu, L., Huang, K., Li, X., Tao, D., et al. (2015). Fibroblast-derived exosomes contribute to chemoresistance through priming cancer stem cells in colorectal cancer. *PLoS One, 10*, e0125625.

Hulleman, E., Kazemier, K. M., Holleman, A., VanderWeele, D. J., Rudin, C. M., Broekhuis, M. J., et al. (2009). Inhibition of glycolysis modulates prednisolone resistance in acute lymphoblastic leukemia cells. *Blood, 113*, 2014–2021.

Iqbal, N., & Iqbal, N. (2014). Imatinib: A breakthrough of targeted therapy in cancer. *Chemotherapy Research and Practice, 2014*, 357027.

Jeon, Y. J., Middleton, J., Kim, T., Lagana, A., Piovan, C., Secchiero, P., et al. (2015). A set of NF-kappaB-regulated microRNAs induces acquired TRAIL resistance in lung cancer. *Proceedings of the National Academy of Sciences of the United States of America, 112*, E3355–64.

Jiang, H., Chen, K., He, J., Pan, F., Li, J., Chen, J., et al. (2009). Association of pregnane X receptor with multidrug resistance-related protein 3 and its role in human colon cancer chemoresistance. *Journal of Gastrointestinal Surgery, 13*, 1831–1838.

Joensuu, H., Pylkkanen, L., & Toikkanen, S. (1994). Bcl-2 protein expression and long-term survival in breast cancer. *The American Journal of Pathology, 145*, 1191–1198.

Johnson, I. S., Armstrong, J. G., Gorman, M., & Burnett, J. P., Jr. (1963). The vinca alkaloids: A new class of oncolytic agents. *Cancer Research, 23*, 1390–1427.

Kang, M. H., & Reynolds, C. P. (2009). Bcl-2 inhibitors: Targeting mitochondrial apoptotic pathways in cancer therapy. *Clinical Cancer Research, 15*, 1126–1132.

Kerbel, R. S., Kobayashi, H., & Graham, C. H. (1994). Intrinsic or acquired drug resistance and metastasis: Are they linked phenotypes? *Journal of Cellular Biochemistry, 56*, 37–47.

Kim, D. W., Kim, K. O., Shin, M. J., Ha, J. H., Seo, S. W., Yang, J., et al. (2009). siRNA-based targeting of antiapoptotic genes can reverse chemoresistance in P-glycoprotein expressing chondrosarcoma cells. *Molecular Cancer, 8*, 28.

Kim, J. H., Sim, S. H., Ha, H. J., Ko, J. J., Lee, K., & Bae, J. (2009). MCL-1ES, a novel variant of MCL-1, associates with MCL-1L and induces mitochondrial cell death. *FEBS Letters, 583*, 2758–2764.

Kischkel, F. C., Hellbardt, S., Behrmann, I., Germer, M., Pawlita, M., Krammer, P. H., et al. (1995). Cytotoxicity-dependent APO-1 (Fas/CD95)-associated proteins form a death-inducing signaling complex (DISC) with the receptor. *The EMBO Journal, 14*, 5579–5588.

Kish, J., Drelichman, A., Jacobs, J., Hoschner, J., Kinzie, J., Loh, J., et al. (1982). Clinical trial of cisplatin and 5-FU infusion as initial treatment for advanced squamous cell carcinoma of the head and neck. *Cancer Treatment Reports, 66*, 471–474.

Klaassen, U., Wilke, H., Muller, C., Borquez, D., Korn, M., Achterrath, W., et al. (1997). Infusional 5-FU, folinic acid, paclitaxel, and cisplatin for metastatic breast cancer. *Oncology, 11*, 38–40.

Konopleva, M., Contractor, R., Tsao, T., Samudio, I., Ruvolo, P. P., Kitada, S., et al. (2006). Mechanisms of apoptosis sensitivity and resistance to the BH3 mimetic ABT-737 in acute myeloid leukemia. *Cancer Cell, 10*, 375–388.

Kucuk, O., Shevrin, D. H., Pandya, K. J., & Bonomi, P. D. (2000). Phase II trial of cisplatin, etoposide, and 5-fluorouracil in advanced non-small-cell lung cancer: An Eastern Cooperative Oncology Group Study (PB586). *American Journal of Clinical Oncology, 23*, 371–375.

Kumar, B., Yadav, A., Lang, J. C., Teknos, T. N., & Kumar, P. (2015). Suberoylanilide hydroxamic acid (SAHA) reverses chemoresistance in head and neck cancer cells by targeting cancer stem cells via the downregulation of nanog. *Genes & Cancer, 6*, 169–181.

Leibowitz, B., & Yu, J. (2010). Mitochondrial signaling in cell death via the Bcl-2 family. *Cancer Biology & Therapy, 9*, 417–422.

Lestini, B. J., Goldsmith, K. C., Fluchel, M. N., Liu, X., Chen, N. L., Goyal, B., et al. (2009). Mcl1 downregulation sensitizes neuroblastoma to cytotoxic chemotherapy and small molecule Bcl2-family antagonists. *Cancer Biology & Therapy, 8*, 1587–1595.

Li, M. C., Hertz, R., & Bergenstal, D. M. (1958). Therapy of choriocarcinoma and related trophoblastic tumors with folic acid and purine antagonists. *The New England Journal of Medicine, 259*, 66–74.

Li, H., Jin, X., Zhang, Z., Xing, Y., & Kong, X. (2013). Inhibition of autophagy enhances apoptosis induced by the PI3K/AKT/mTor inhibitor NVP-BEZ235 in renal cell carcinoma cells. *Cell Biochemistry and Function, 31*, 427–433.

Liao, M., Zhao, J., Wang, T., Duan, J., Zhang, Y., & Deng, X. (2011). Role of bile salt in regulating Mcl-1 phosphorylation and chemoresistance in hepatocellular carcinoma cells. *Molecular Cancer, 10*, 44.

Lin, K. H., Winter, P. S., Xie, A., Roth, C., Martz, C. A., Stein, E. M., et al. (2016). Targeting MCL-1/BCL-XL forestalls the acquisition of resistance to ABT-199 in acute myeloid leukemia. *Scientific Reports, 6*, 27696.

Linde, N., Fluegen, G., & Aguirre-Ghiso, J. A. (2016). The relationship between dormant cancer cells and their microenvironment. *Advances in Cancer Research, 132*, 45–71.

Linn, S. C., & Giaccone, G. (1995). MDR1/P-glycoprotein expression in colorectal cancer. *European Journal of Cancer, 31A*, 1291–1294.

Liu, Y., Cao, Y., Zhang, W., Bergmeier, S., Qian, Y., Akbar, H., et al. (2012). A small-molecule inhibitor of glucose transporter 1 downregulates glycolysis, induces cell-cycle arrest, and inhibits cancer cell growth in vitro and in vivo. *Molecular Cancer Therapeutics, 11*, 1672–1682.

Liu, J., & Debnath, J. (2016). The evolving, multifaceted roles of autophagy in cancer. *Advances in Cancer Research, 130*, 1–53.

Liu, Y., Gao, S., Chen, X., Liu, M., Mao, C., & Fang, X. (2016). Overexpression of miR-203 sensitizes paclitaxel (Taxol)-resistant colorectal cancer cells through targeting the salt-inducible kinase 2 (SIK2). *Tumour Biology: The Journal of the International Society for Oncodevelopmental Biology and Medicine, 37*, 12231–12239.

Liu, C., Li, Z., Bi, L., Li, K., Zhou, B., Xu, C., et al. (2014). NOTCH1 signaling promotes chemoresistance via regulating ABCC1 expression in prostate cancer stem cells. *Molecular and Cellular Biochemistry, 393*, 265–270.

Liu, H., Liu, Y., & Zhang, J. T. (2008). A new mechanism of drug resistance in breast cancer cells: Fatty acid synthase overexpression-mediated palmitate overproduction. *Molecular Cancer Therapeutics, 7*, 263–270.

Liu, M. X., Siu, M. K., Liu, S. S., Yam, J. W., Ngan, H. Y., & Chan, D. W. (2014). Epigenetic silencing of microRNA-199b-5p is associated with acquired chemoresistance via activation of JAG1-Notch1 signaling in ovarian cancer. *Oncotarget, 5*, 944–958.

Liu, T., Xu, F., Du, X., Lai, D., Liu, T., Zhao, Y., et al. (2010). Establishment and characterization of multi-drug resistant, prostate carcinoma-initiating stem-like cells from human prostate cancer cell lines 22RV1. *Molecular and Cellular Biochemistry, 340*, 265–273.

Mahon, F. X., Belloc, F., Lagarde, V., Chollet, C., Moreau-Gaudry, F., Reiffers, J., et al. (2003). MDR1 gene overexpression confers resistance to imatinib mesylate in leukemia cell line models. *Blood, 101*, 2368–2373.

Maji, S., Samal, S. K., Pattanaik, L., Panda, S., Quinn, B. A., Das, S. K., et al. (2015). Mcl-1 is an important therapeutic target for oral squamous cell carcinomas. *Oncotarget, 6*, 16623–16637.

Mayer, M. J., Klotz, L. H., & Venkateswaran, V. (2015). Metformin and prostate cancer stem cells: A novel therapeutic target. *Prostate Cancer and Prostatic Diseases, 18*, 303–309.

McBrayer, S. K., Cheng, J. C., Singhal, S., Krett, N. L., Rosen, S. T., & Shanmugam, M. (2012). Multiple myeloma exhibits novel dependence on GLUT4, GLUT8, and GLUT11: Implications for glucose transporter-directed therapy. *Blood, 119*, 4686–4697.

McDonnell, T. J., & Korsmeyer, S. J. (1991). Progression from lymphoid hyperplasia to high-grade malignant lymphoma in mice transgenic for the t(14; 18). *Nature, 349*, 254–256.

McDonnell, T. J., Troncoso, P., Brisbay, S. M., Logothetis, C., Chung, L. W., Hsieh, J. T., et al. (1992). Expression of the protooncogene bcl-2 in the prostate and its association with emergence of androgen-independent prostate cancer. *Cancer Research, 52*, 6940–6944.

McIntosh, K., Balch, C., & Tiwari, A. K. (2016). Tackling multidrug resistance mediated by efflux transporters in tumor-initiating cells. *Expert Opinion on Drug Metabolism & Toxicology, 12*, 633–644.

Menendez, J. A., Lupu, R., & Colomer, R. (2004). Inhibition of tumor-associated fatty acid synthase hyperactivity induces synergistic chemosensitization of HER-2/neu-overexpressing human breast cancer cells to docetaxel (taxotere). *Breast Cancer Research and Treatment, 84*, 183–195.

Menendez, J. A., Vellon, L., & Lupu, R. (2005). Targeting fatty acid synthase-driven lipid rafts: A novel strategy to overcome trastuzumab resistance in breast cancer cells. *Medical Hypotheses, 64*, 997–1001.

Meynet, O., Beneteau, M., Jacquin, M. A., Pradelli, L. A., Cornille, A., Carles, M., et al. (2012). Glycolysis inhibition targets Mcl-1 to restore sensitivity of lymphoma cells to ABT-737-induced apoptosis. *Leukemia, 26*, 1145–1147.

Michels, J., Obrist, F., Vitale, I., Lissa, D., Garcia, P., Behnam-Motlagh, P., et al. (2014). MCL-1 dependency of cisplatin-resistant cancer cells. *Biochemical Pharmacology, 92*, 55–61.

Miyashita, T., & Reed, J. C. (1993). Bcl-2 oncoprotein blocks chemotherapy-induced apoptosis in a human leukemia cell line. *Blood, 81*, 151–157.

Moitra, K. (2015). Overcoming multidrug resistance in cancer stem cells. *BioMed Research International, 2015*, 635745.

Morales, A. A., Olsson, A., Celsing, F., Osterborg, A., Jondal, M., & Osorio, L. M. (2005). High expression of bfl-1 contributes to the apoptosis resistant phenotype in B-cell chronic lymphocytic leukemia. *International Journal of Cancer, 113*, 730–737.

Morgan, R. J., Jr., Braly, P., Leong, L., Shibata, S., Margolin, K., Somlo, G., et al. (2000). Phase II trial of combination intraperitoneal cisplatin and 5-fluorouracil in previously treated patients with advanced ovarian cancer: Long-term follow-up. *Gynecologic Oncology, 77*, 433–438.

Munoz-Pinedo, C., El Mjiyad, N., & Ricci, J. E. (2012). Cancer metabolism: Current perspectives and future directions. *Cell Death & Disease, 3*, e248.

Naik, E., Michalak, E. M., Villunger, A., Adams, J. M., & Strasser, A. (2007). Ultraviolet radiation triggers apoptosis of fibroblasts and skin keratinocytes mainly via the BH3-only protein Noxa. *The Journal of Cell Biology, 176*, 415–424.

Ni, J., Cozzi, P., Hao, J., Duan, W., Graham, P., Kearsley, J., et al. (2014). Cancer stem cells in prostate cancer chemoresistance. *Current Cancer Drug Targets, 14*, 225–240.

Nishioka, T., Luo, L. Y., Shen, L., He, H., Mariyannis, A., Dai, W., et al. (2014). Nicotine increases the resistance of lung cancer cells to cisplatin through enhancing Bcl-2 stability. *British Journal of Cancer, 110*, 1785–1792.

Oh, W. K., George, D. J., & Tay, M. H. (2005). Response to docetaxel/carboplatin in patients with hormone-refractory prostate cancer not responding to taxane-based chemotherapy. *Clinical Prostate Cancer, 4*, 61–64.

Okamoto, K., Miyoshi, K., & Murawaki, Y. (2013). miR-29b, miR-205 and miR-221 enhance chemosensitivity to gemcitabine in HuH28 human cholangiocarcinoma cells. *PLoS One, 8,* e77623.

Olsson, A., Norberg, M., Okvist, A., Derkow, K., Choudhury, A., Tobin, G., et al. (2007). Upregulation of bfl-1 is a potential mechanism of chemoresistance in B-cell chronic lymphocytic leukaemia. *British Journal of Cancer, 97,* 769–777.

Osaki, S., Tazawa, H., Hasei, J., Yamakawa, Y., Omori, T., Sugiu, K., et al. (2016). Ablation of MCL1 expression by virally induced microRNA-29 reverses chemoresistance in human osteosarcomas. *Scientific Reports, 6,* 28953.

Pan, R., Hogdal, L. J., Benito, J. M., Bucci, D., Han, L., Borthakur, G., et al. (2014). Selective BCL-2 inhibition by ABT-199 causes on-target cell death in acute myeloid leukemia. *Cancer Discovery, 4,* 362–375.

Park, I. C., Lee, S. H., Whang, D. Y., Hong, W. S., Choi, S. S., Shin, H. S., et al. (1997). Expression of a novel Bcl-2 related gene, Bfl-1, in various human cancers and cancer cell lines. *Anticancer Research, 17,* 4619–4622.

Placzek, W. J., Wei, J., Kitada, S., Zhai, D., Reed, J. C., & Pellecchia, M. (2010). A survey of the anti-apoptotic Bcl-2 subfamily expression in cancer types provides a platform to predict the efficacy of Bcl-2 antagonists in cancer therapy. *Cell Death & Disease, 1,* e40.

Quinn, B. A., Dash, R., Azab, B., Sarkar, S., Das, S. K., Kumar, S., et al. (2011). Targeting Mcl-1 for the therapy of cancer. *Expert Opinion on Investigational Drugs, 20,* 1397–1411.

Rahman, M., & Hasan, M. R. (2015). Cancer metabolism and drug resistance. *Metabolites, 5,* 571–600.

Ronnekleiv-Kelly, S. M., Sharma, A., & Ahuja, N. (2017). Epigenetic therapy and chemosensitization in solid malignancy. *Cancer Treatment Reviews, 55,* 200–208.

Rosell, R., Gatzemeier, U., Betticher, D. C., Keppler, U., Macha, H. N., Pirker, R., et al. (2002). Phase III randomised trial comparing paclitaxel/carboplatin with paclitaxel/cisplatin in patients with advanced non-small-cell lung cancer: A cooperative multinational trial. *Annals of Oncology, 13,* 1539–1549.

Rosenberg, B., Vancamp, L., & Krigas, T. (1965). Inhibition of cell division in Escherichia coli by electrolysis products from a platinum electrode. *Nature, 205,* 698–699.

Sabnis, N. G., Miller, A., Titus, M. A., & Huss, W. J. (2017). The efflux transporter ABCG2 maintains prostate stem cells. *Molecular Cancer Research: MCR, 15,* 128–140.

Saha, S., Mukherjee, S., Khan, P., Kajal, K., Mazumdar, M., Manna, A., et al. (2016). Aspirin suppresses the acquisition of chemoresistance in breast cancer by disrupting an NFkappaB-IL6 signaling axis responsible for the generation of cancer stem cells. *Cancer Research, 76,* 2000–2012.

Sakamoto, T., Kobayashi, S., Yamada, D., Nagano, H., Tomokuni, A., Tomimaru, Y., et al. (2016). A histone deacetylase inhibitor suppresses epithelial-mesenchymal transition and attenuates chemoresistance in biliary tract cancer. *PLoS One, 11,* e0145985.

Saleem, A., Dvorzhinski, D., Santanam, U., Mathew, R., Bray, K., Stein, M., et al. (2012). Effect of dual inhibition of apoptosis and autophagy in prostate cancer. *The Prostate, 72,* 1374–1381.

Sartorius, U. A., & Krammer, P. H. (2002). Upregulation of Bcl-2 is involved in the mediation of chemotherapy resistance in human small cell lung cancer cell lines. *International Journal of Cancer, 97,* 584–592.

Sasaki, K., Tsuno, N. H., Sunami, E., Tsurita, G., Kawai, K., Okaji, Y., et al. (2010). Chloroquine potentiates the anti-cancer effect of 5-fluorouracil on colon cancer cells. *BMC Cancer, 10,* 370.

Schmitt, C. A., & Lowe, S. W. (2001). Bcl-2 mediates chemoresistance in matched pairs of primary E(mu)-myc lymphomas in vivo. *Blood Cells, Molecules & Diseases, 27,* 206–216.

Schneider, J., Gonzalez-Roces, S., Pollan, M., Lucas, R., Tejerina, A., Martin, M., et al. (2001). Expression of LRP and MDR1 in locally advanced breast cancer predicts axillary node invasion at the time of rescue mastectomy after induction chemotherapy. *Breast Cancer Research: BCR, 3*, 183–191.

Schoenlein, P. V., Periyasamy-Thandavan, S., Samaddar, J. S., Jackson, W. H., & Barrett, J. T. (2009). Autophagy facilitates the progression of ERalpha-positive breast cancer cells to antiestrogen resistance. *Autophagy, 5*, 400–403.

Schwickart, M., Huang, X., Lill, J. R., Liu, J., Ferrando, R., French, D. M., et al. (2010). Deubiquitinase USP9X stabilizes MCL1 and promotes tumour cell survival. *Nature, 463*, 103–107.

Shafee, N., Smith, C. R., Wei, S., Kim, Y., Mills, G. B., Hortobagyi, G. N., et al. (2008). Cancer stem cells contribute to cisplatin resistance in Brca1/p53-mediated mouse mammary tumors. *Cancer Research, 68*, 3243–3250.

Shintani, T., & Klionsky, D. J. (2004). Autophagy in health and disease: A double-edged sword. *Science, 306*, 990–995.

Shintani, Y., Okimura, A., Sato, K., Nakagiri, T., Kadota, Y., Inoue, M., et al. (2011). Epithelial to mesenchymal transition is a determinant of sensitivity to chemoradiotherapy in non-small cell lung cancer. *The Annals of Thoracic Surgery, 92*, 1794–1804. discussion 1804.

Siegel, R. L., Miller, K. D., & Jemal, A. (2017). Cancer statistics, 2017. *CA: A Cancer Journal for Clinicians, 67*, 7–30.

Sieghart, W., Losert, D., Strommer, S., Cejka, D., Schmid, K., Rasoul-Rockenschaub, S., et al. (2006). Mcl-1 overexpression in hepatocellular carcinoma: A potential target for antisense therapy. *Journal of Hepatology, 44*, 151–157.

Skipper, H. E., Thomson, J. R., Elion, G. B., & Hitchings, G. H. (1954). Observations on the anticancer activity of 6-mercaptopurine. *Cancer Research, 14*, 294–298.

Skubisz, M. M., & Tong, S. (2012). The evolution of methotrexate as a treatment for ectopic pregnancy and gestational trophoblastic neoplasia: A review. *ISRN Obstetrics and Gynecology, 2012*, 637094.

Song, B., Wang, Y., Titmus, M. A., Botchkina, G., Formentini, A., Kornmann, M., et al. (2010). Molecular mechanism of chemoresistance by miR-215 in osteosarcoma and colon cancer cells. *Molecular Cancer, 9*, 96.

Song, B., Wang, Y., Xi, Y., Kudo, K., Bruheim, S., Botchkina, G. I., et al. (2009). Mechanism of chemoresistance mediated by miR-140 in human osteosarcoma and colon cancer cells. *Oncogene, 28*, 4065–4074.

Stacy, A. E., Jansson, P. J., & Richardson, D. R. (2013). Molecular pharmacology of ABCG2 and its role in chemoresistance. *Molecular Pharmacology, 84*, 655–669.

Stanton, M. J., Dutta, S., Zhang, H., Polavaram, N. S., Leontovich, A. A., Honscheid, P., et al. (2013). Autophagy control by the VEGF-C/NRP-2 axis in cancer and its implication for treatment resistance. *Cancer Research, 73*, 160–171.

Steg, A. D., Bevis, K. S., Katre, A. A., Ziebarth, A., Dobbin, Z. C., Alvarez, R. D., et al. (2012). Stem cell pathways contribute to clinical chemoresistance in ovarian cancer. *Clinical Cancer Research, 18*, 869–881.

Strik, H., Deininger, M., Streffer, J., Grote, E., Wickboldt, J., Dichgans, J., et al. (1999). BCL-2 family protein expression in initial and recurrent glioblastomas: Modulation by radiochemotherapy. *Journal of Neurology, Neurosurgery, & Psychiatry, 67*, 763–768.

Sui, X., Chen, R., Wang, Z., Huang, Z., Kong, N., Zhang, M., et al. (2013). Autophagy and chemotherapy resistance: A promising therapeutic target for cancer treatment. *Cell Death & Disease, 4*, e838.

Sun, Y. L., Chen, J. J., Kumar, P., Chen, K., Sodani, K., Patel, A., et al. (2013). Reversal of MRP7 (ABCC10)-mediated multidrug resistance by tariquidar. *PLoS One, 8*, e55576.

Tabuchi, Y., Matsuoka, J., Gunduz, M., Imada, T., Ono, R., Ito, M., et al. (2009). Resistance to paclitaxel therapy is related with Bcl-2 expression through an estrogen receptor mediated pathway in breast cancer. *International Journal of Oncology, 34*, 313–319.

Talukdar, S., Emdad, L., Das, S. K., Sarkar, D., & Fisher, P. B. (2016). Evolving strategies for therapeutically targeting cancer stem cells. *Advances in Cancer Research, 131*, 159–191.

Tao, L., Huang, G., Chen, Y., & Chen, L. (2015). DNA methylation of DKK3 modulates docetaxel chemoresistance in human nonsmall cell lung cancer cell. *Cancer Biotherapy & Radiopharmaceuticals, 30*, 100–106.

Thomas, S., Quinn, B. A., Das, S. K., Dash, R., Emdad, L., Dasgupta, S., et al. (2013). Targeting the Bcl-2 family for cancer therapy. *Expert Opinion on Therapeutic Targets, 17*, 61–75.

Toge, M., Yokoyama, S., Kato, S., Sakurai, H., Senda, K., Doki, Y., et al. (2015). Critical contribution of MCL-1 in EMT-associated chemo-resistance in A549 non-small cell lung cancer. *International Journal of Oncology, 46*, 1844–1848.

Triller, N., Korosec, P., Kern, I., Kosnik, M., & Debeljak, A. (2006). Multidrug resistance in small cell lung cancer: Expression of P-glycoprotein, multidrug resistance protein 1 and lung resistance protein in chemo-naive patients and in relapsed disease. *Lung Cancer, 54*, 235–240.

Varin, E., Denoyelle, C., Brotin, E., Meryet-Figuiere, M., Giffard, F., Abeilard, E., et al. (2010). Downregulation of Bcl-xL and Mcl-1 is sufficient to induce cell death in mesothelioma cells highly refractory to conventional chemotherapy. *Carcinogenesis, 31*, 984–993.

Vazquez-Martin, A., Ropero, S., Brunet, J., Colomer, R., & Menendez, J. A. (2007). Inhibition of fatty acid synthase (FASN) synergistically enhances the efficacy of 5-fluorouracil in breast carcinoma cells. *Oncology Reports, 18*, 973–980.

Vidal, S. J., Rodriguez-Bravo, V., Galsky, M., Cordon-Cardo, C., & Domingo-Domenech, J. (2014). Targeting cancer stem cells to suppress acquired chemotherapy resistance. *Oncogene, 33*, 4451–4463.

Vogler, M. (2012). BCL2A1: The underdog in the BCL2 family. *Cell Death and Differentiation, 19*, 67–74.

Vogler, M., Butterworth, M., Majid, A., Walewska, R. J., Sun, X. M., Dyer, M. J., et al. (2009). Concurrent up-regulation of BCL-XL and BCL2A1 induces approximately 1000-fold resistance to ABT-737 in chronic lymphocytic leukemia. *Blood, 113*, 4403–4413.

Wang, L., Liu, X., Ren, Y., Zhang, J., Chen, J., Zhou, W., et al. (2017). Cisplatin-enriching cancer stem cells confer multidrug resistance in non-small cell lung cancer via enhancing TRIB1/HDAC activity. *Cell Death & Disease, 8*, e2746.

Wang, L., Mosel, A. J., Oakley, G. G., & Peng, A. (2012). Deficient DNA damage signaling leads to chemoresistance to cisplatin in oral cancer. *Molecular Cancer Therapeutics, 11*, 2401–2409.

Wasserheit, C., Frazein, A., Oratz, R., Sorich, J., Downey, A., Hochster, H., et al. (1996). Phase II trial of paclitaxel and cisplatin in women with advanced breast cancer: An active regimen with limiting neurotoxicity. *Journal of Clinical Oncology, 14*, 1993–1999.

Williams, J., Lucas, P. C., Griffith, K. A., Choi, M., Fogoros, S., Hu, Y. Y., et al. (2005). Expression of Bcl-xL in ovarian carcinoma is associated with chemoresistance and recurrent disease. *Gynecologic Oncology, 96*, 287–295.

Wu, C. P., Calcagno, A. M., & Ambudkar, S. V. (2008). Reversal of ABC drug transporter-mediated multidrug resistance in cancer cells: Evaluation of current strategies. *Current Molecular Pharmacology, 1*, 93–105.

Xie, K., Doles, J., Hemann, M. T., & Walker, G. C. (2010). Error-prone translesion synthesis mediates acquired chemoresistance. *Proceedings of the National Academy of Sciences of the United States of America, 107*, 20792–20797.

Yamaguchi, R., Janssen, E., Perkins, G., Ellisman, M., Kitada, S., & Reed, J. C. (2011). Efficient elimination of cancer cells by deoxyglucose-ABT-263/737 combination therapy. *PLoS One, 6,* e24102.

Yang, S., Irani, K., Heffron, S. E., Jurnak, F., & Meyskens, F. L., Jr. (2005). Alterations in the expression of the apurinic/apyrimidinic endonuclease-1/redox factor-1 (APE/Ref-1) in human melanoma and identification of the therapeutic potential of resveratrol as an APE/Ref-1 inhibitor. *Molecular Cancer Therapeutics, 4,* 1923–1935.

Yang, N., Kaur, S., Volinia, S., Greshock, J., Lassus, H., Hasegawa, K., et al. (2008). MicroRNA microarray identifies Let-7i as a novel biomarker and therapeutic target in human epithelial ovarian cancer. *Cancer Research, 68,* 10307–10314.

Yang, Y., Liu, H., Li, Z., Zhao, Z., Yip-Schneider, M., Fan, Q., et al. (2011). Role of fatty acid synthase in gemcitabine and radiation resistance of pancreatic cancers. *International Journal of Biochemistry and Molecular Biology, 2,* 89–98.

Yang, J., Zhang, K., Wu, J., Shi, J., Xue, J., Li, J., et al. (2016). Wnt5a increases properties of lung cancer stem cells and resistance to cisplatin through activation of Wnt5a/PKC signaling pathway. *Stem Cells International, 2016,* 1690896.

Yang, X. K., Zheng, F., Chen, J. H., Gao, Q. L., Lu, Y. P., Wang, S. X., et al. (2002). Relationship between expression of apoptosis-associated proteins and caspase-3 activity in cisplatin-resistant human ovarian cancer cell line. *Chinese Journal of Cancer, 21,* 1288–1291.

Yecies, D., Carlson, N. E., Deng, J., & Letai, A. (2010). Acquired resistance to ABT-737 in lymphoma cells that up-regulate MCL-1 and Bfl-1. *Blood, 115,* 3304–3313.

Youle, R. J., & Strasser, A. (2008). The BCL-2 protein family: Opposing activities that mediate cell death. *Nature Reviews Molecular Cell Biology, 9,* 47–59.

Zalcberg, J., Hu, X. F., Slater, A., Parisot, J., El-Osta, S., Kantharidis, P., et al. (2000). MRP1 not MDR1 gene expression is the predominant mechanism of acquired multidrug resistance in two prostate carcinoma cell lines. *Prostate Cancer and Prostatic Diseases, 3,* 66–75.

Zarogoulidis, P., Petanidis, S., Kioseoglou, E., Domvri, K., Anestakis, D., & Zarogoulidis, K. (2015). MiR-205 and miR-218 expression is associated with carboplatin chemoresistance and regulation of apoptosis via Mcl-1 and Survivin in lung cancer cells. *Cellular Signalling, 27,* 1576–1588.

Zhang, Y., Cheng, Y., Ren, X., Zhang, L., Yap, K. L., Wu, H., et al. (2012). NAC1 modulates sensitivity of ovarian cancer cells to cisplatin by altering the HMGB1-mediated autophagic response. *Oncogene, 31,* 1055–1064.

Zhang, N., Qi, Y., Wadham, C., Wang, L., Warren, A., Di, W., et al. (2010). FTY720 induces necrotic cell death and autophagy in ovarian cancer cells: A protective role of autophagy. *Autophagy, 6,* 1157–1167.

Zhang, Y., Xu, W., Guo, H., Zhang, Y., He, Y., Lee, S. H., et al. (2017). NOTCH1 signaling regulates self-renewal and platinum chemoresistance of cancer stem-like cells in human non-small cell lung cancer. *Cancer Research, 77,* 3082–3091.

Zhao, Q., Cao, J., Wu, Y. C., Liu, X., Han, J., Huang, X. C., et al. (2015). Circulating miRNAs is a potential marker for gefitinib sensitivity and correlation with EGFR mutational status in human lung cancers. *American Journal of Cancer Research, 5,* 1692–1705.

Zhao, Y., Liu, H., Liu, Z., Ding, Y., Ledoux, S. P., Wilson, G. L., et al. (2011). Overcoming trastuzumab resistance in breast cancer by targeting dysregulated glucose metabolism. *Cancer Research, 71,* 4585–4597.

Zhao, Y., Lu, H., Yan, A., Yang, Y., Meng, Q., Sun, L., et al. (2013). ABCC3 as a marker for multidrug resistance in non-small cell lung cancer. *Scientific Reports, 3,* 3120.

Zhao, L., Ren, Y., Tang, H., Wang, W., He, Q., Sun, J., et al. (2015). Deregulation of the miR-222-ABCG2 regulatory module in tongue squamous cell carcinoma contributes to chemoresistance and enhanced migratory/invasive potential. *Oncotarget, 6,* 44538–44550.

Zhao, Y., Zhang, C. L., Zeng, B. F., Wu, X. S., Gao, T. T., & Oda, Y. (2009). Enhanced chemosensitivity of drug-resistant osteosarcoma cells by lentivirus-mediated Bcl-2 silencing. *Biochemical and Biophysical Research Communications, 390*, 642–647.

Zhou, J. J., Deng, X. G., He, X. Y., Zhou, Y., Yu, M., Gao, W. C., et al. (2014). Knockdown of NANOG enhances chemosensitivity of liver cancer cells to doxorubicin by reducing MDR1 expression. *International Journal of Oncology, 44*, 2034–2040.

Zhou, P., Qian, L., Kozopas, K. M., & Craig, R. W. (1997). Mcl-1, a Bcl-2 family member, delays the death of hematopoietic cells under a variety of apoptosis-inducing conditions. *Blood, 89*, 630–643.

Zhou, M., Zhao, Y., Ding, Y., Liu, H., Liu, Z., Fodstad, O., et al. (2010). Warburg effect in chemosensitivity: Targeting lactate dehydrogenase-A re-sensitizes taxol-resistant cancer cells to taxol. *Molecular Cancer, 9*, 33.

CHAPTER FOUR

# New Insights Into Beclin-1: Evolution and Pan-Malignancy Inhibitor Activity

Stephen L. Wechman[*], Anjan K. Pradhan[*], Rob DeSalle[§],
Swadesh K. Das[*,†,‡], Luni Emdad[*,†,‡], Devanand Sarkar[*,†,‡],
Paul B. Fisher[*,†,‡,1]

[*]Virginia Commonwealth University, School of Medicine, Richmond, VA, United States
[†]VCU Institute of Molecular Medicine, Virginia Commonwealth University, School of Medicine, Richmond, VA, United States
[‡]VCU Massey Cancer Center, Virginia Commonwealth University, School of Medicine, Richmond, VA, United States
[§]Sackler Institute for Comparative Genomics, American Museum of Natural History, New York, NY, United States
[1]Corresponding author e-mail address: paul.fisher@vcuhealth.org

## Contents

| | |
|---|---|
| 1. General Introduction | 78 |
| 2. Introduction to Beclin-1 | 79 |
|    2.1 Molecular Evolution of Eukaryotic Beclin-1 | 82 |
|    2.2 Beclin-1 and Eukaryotic Cell Autophagy | 84 |
| 3. Beclin-1 and Cancer | 85 |
|    3.1 Beclin-1 as a Tumor Suppressor | 86 |
|    3.2 The *mda-7*-miR-221-Beclin-1 Axis | 90 |
|    3.3 Arguments for and Against the Oncogenicity of Beclin-1 | 92 |
|    3.4 Beclin-1 Expression and Its Effect Upon Cancer Progression | 97 |
|    3.5 Beclin-1 and Resistance to Cancer Therapy | 99 |
|    3.6 Beclin-1 and CSC Maintenance | 102 |
| 4. Conclusions and Future Directions | 103 |
| Acknowledgments | 103 |
| Supplementary Material | 104 |
| References | 104 |

## Abstract

Autophagy is a functionally conserved self-degradation process that facilitates the survival of eukaryotic life via the management of cellular bioenergetics and maintenance of the fidelity of genomic DNA. The first known autophagy inducer was Beclin-1. Beclin-1 is expressed in multicellular eukaryotes ranging throughout plants to animals, comprising a nonmonophyllic group, as shown in this report via aggressive BLAST searches. In humans, Beclin-1 is a haploinsuffient tumor suppressor as biallelic deletions have not been observed in patient tumors clinically. Therefore, Beclin-1 fails the Knudson hypothesis, implicating expression of at least one Beclin-1 allele is essential for cancer cell survival. However, Beclin-1 is frequently monoallelically deleted in advanced human cancers and the expression of two Beclin-1 allelles is associated with greater anticancer effects. Overall, experimental evidence suggests that Beclin-1 inhibits tumor formation, angiogenesis, and metastasis alone and in cooperation with the tumor suppressive molecules UVRAG, Bif-1, Ambra1, and MDA-7/IL-24 via diverse mechanisms of action. Conversely, Beclin-1 is upregulated in cancer stem cells (CSCs), portending a role in cancer recurrence, and highlighting this molecule as an intriguing molecular target for the treatment of CSCs. Many aspects of Beclin-1's biological effects remain to be studied. The consequences of these BLAST searches on the molecular evolution of Beclin-1, and the eukaryotic branches of the tree of life, are discussed here in greater detail with future inquiry focused upon protist taxa. Also in this review, the effects of Beclin-1 on tumor suppression and cancer malignancy are discussed. Beclin-1 holds significant promise for the development of novel targeted cancer therapeutics and is anticipated to lead to a many advances in our understanding of eukaryotic evolution, multicellularity, and even the treatment of CSCs in the coming decades.

## 1. GENERAL INTRODUCTION

Autophagy remains a topic of intensive investigation due to its affect upon cancer including cell survival, invasion, and metastasis; or conversely, the suppression of tumor growth, through lysing cells via type II programmed cell death (PCD; Kroemer et al., 2009). These distinct affects are thought to vary depending on the tissue, redox status, and other biochemical pathways which are activated concurrently with autophagy signaling. Generally, autophagy leads to the sequestration of organelles and long-lived proteins into autophagosomes, which later fuse with lysosomes to form the autophagolysosome, degrading those sequestered cellular components to produce energy and cellular monomers (Fig. 1; Ren & Taegtmeyer, 2015). The maintenance of autophagy flux is essential to successfully accomplish mammalian embryogenesis. As such, autophagy is evolutionarily conserved in eukaryotes, supporting the significance of this biochemical pathway. The term "autophagy" can refer to many different subclassifications of this process; for the purpose of this review, "autophagy" refers to macroautophagy.

As many as 30 autophagy-related genes (Atg) have been identified in *Saccharomyces cerevisiae*; please refer to Cao et al. and Klionsky et al. for a more complete review of these genes and their molecular mechanisms

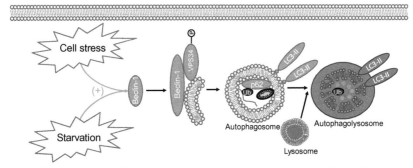

**Fig. 1** The effect of cell stress and starvation on the maturation of the autophagolysosomes and autophagy. This graphic depicts the extension of the autophagy membrane, the sequestration of ubiquitinated long-lived proteins, p62-mediated protein scavenging, and damaged organelles such as the mitochondria. Ultimately, these components are degraded following lysosomal fusion. *Blue circles* are representative of degraded proteins and *orange circles* are representative of the degradation of the mitochondria. *Purple coloring* is indicative of autophagosome acidification and the degradation of its components occurring during the final stages of autophagolysosome maturation.

(Cao & Klionsky, 2007; Li et al., 2012). Beclin-1 was initially cloned in 1998 (Liang et al., 1998) and was the first identified mammalian autophagy inducing protein (Liang et al., 1999). Beclin-1 has now been shown to participate in many other biological processes including gametogenesis, neurodegeneration, apoptosis, and tumorigenesis (Mizushima, Levine, Cuervo, & Klionsky, 2008). Despite the well-characterized interaction of Beclin-1 and cancer, no prior review has comprehensively discussed the evolution and cancer therapeutic functions of Beclin-1, to our knowledge, at this time. Accordingly, to better understand the cancer therapeutic effects of Beclin-1, this review will discuss: (1) the molecular evolution of Beclin-1, (2) Beclin-1's tumor suppressive mechanisms of action, (3) the effects of Beclin-1 expression upon tumors and cancer malignancy, (4) the effect of Beclin-1 upon conventional cancer therapeutics, and (5) the effect of Beclin-1 upon cancer stem cells (CSCs).

## 2. INTRODUCTION TO BECLIN-1

Following the discovery and characterization of Beclin-1, DNA sequencing analysis revealed that Beclin-1 expression is conserved in all animals and most other multicellular eukaryotes (Cuervo, 2004; Klionsky, 2007; Levine & Klionsky, 2004; Mizushima, 2007; Mizushima et al., 2008), indicating the significance of Beclin-1 at the earliest stages of

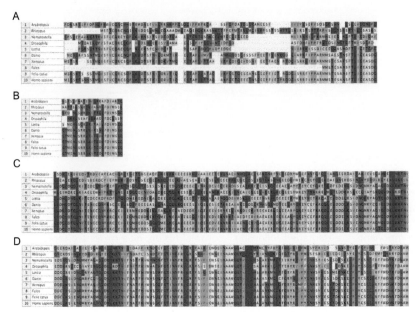

**Fig. 2** The conservation of Beclin-1 amino acid sequences between humans and 10 critical multicellular eukaryotic taxa. (A) The alignment of Beclin-1 5′ amino acids (1–113). (B) Alignment of the Beclin-1 Bcl-2 homology domain 3 (BH3; 114–130) amino acids. (C) Alignment of the Beclin-1 CCD region (144–269) amino acids. (D) The alignment of the evolutionarily conserved domain (ECD; 244–337 aa's). The numbers 1–10 are indicative of the following taxa, respectively. 1 = *Arabidopsis* (plant), 2 = *Rhizopus* (fungi), 3 = *Nematostella* (Cnidaria), 4 = *Drosophila* (Protostome—Ecdysozoa), 5 = *Lottia* (Protostome—Lophotrochozoa), 6 = *Danio* (fish), 7 = *Xenopus* (amphibian), 8 = *Falco* (bird), 9 = *Felis* (cat), and 10 = *Homo* (human). BH3, Bcl-2 homology domain; CCD, coiled-coil domain; ECD, evolutionarily conserved domain.

eukaryotic divergence (Fig. 2). Interestingly, Beclin-1 preforms nearly identical functions over the diversity of animals, promoting their survival in response to stress via autophagy to degrade damaged cellular proteins and organelles. In this way, Beclin-1 plays a critical role in maintaining the complex balance of cellular bioenergetics during periods of stress and starvation in a wide range of organisms. Furthermore, Beclin-1 has also been shown to play a role during embryogenesis and reproduction of eukaryotic organisms ranging from plants (Singh et al., 2010) to mice (Gawriluk, Ko, Hong, Christenson, & Rucker, 2014). Therefore, Beclin-1 appears to be a key regulator of not only survival, but also the life cycle of most multicellular eukaryotes.

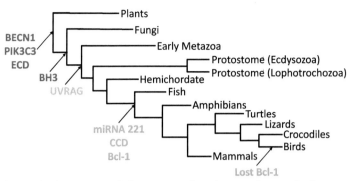

**Fig. 3** The evolutionary tree of the origins of Beclin-1 and several of its interaction partners in eukaryotes. Specific protein domains of BECN1 are also indicated. BECN1, Beclin-1; ECD, evolutionarily conserved domain; BH3, Bcl-2 homology domain; UVRAG, ultraviolet radiation resistance-associated gene; CCD, coiled-coil domain.

Beclin-1 contains many domains with conserved amino acid sequences across eukaryotic species (Fig. 3). One of these evolutionarily conserved domains (ECD; 244–377 aa's) is essential for mediating the protein–protein interaction of Beclin-1 and the class III phosphatidylinositol 3′-kinase (PI3K) Vps34, facilitating eukaryotic cell adaptation to stressful conditions (Furuya, Yu, Byfield, Pattingre, & Levine, 2005). In this manner, it is likely that Beclin-1 provided a selective advantage during the early stages of eukaryotic evolution as primitive eukaryotes biochemically diverged from prokaryotes, however, the point of this divergence remains unknown. For example, Beclin-1 has been shown to contribute to DNA stability, enhance DNA repair, and strengthen overall DNA genome integrity via the interaction with Topoisomerase-IIβ (Xu et al., 2017). For this interaction to occur, Beclin-1 must translocate to the nucleus. The nuclear translocation of Beclin-1 appears to occur early during murine development (15–20 days postnatal; Xu et al., 2017). As mammals, such as mice, mature Beclin-1 translocates back to the cytoplasm while some of these proteins (~50%) remain in the nucleus. However, Beclin-1 and Topoisomerase-IIβ are not the only DNA damage repair proteins expressed in eukaryotic cells. Other examples include heat shock proteins (HSPs), which are expressed in both prokaryotic and eukaryotic cells as an adaptation to cell stress (Finka & Goloubinoff, 2013). It is intriguing, however, that Beclin-1 appears to have been selectively advantageous to eukaryotic, but not prokaryotic cells. Furthermore, if Beclin-1 provided an advantage to prokaryotic cells, then it would have undoubtedly arisen in the prokaryotic cell lineage, indicating that there are unique aspects of eukaryotic cell biochemistry which led to the molecular evolution of Beclin-1. Some of these unique eukaryotic cell factors may have included the development of the nucleus and the longer-life spans observed in eukaryotic cells. It is possible that the expression of the HSPs alone were not sufficient for multicellular eukaryotes and additional biochemical pathways, such as Beclin-1, were necessary to ensure the fidelity of the eukaryotic DNA genome and cell viability in response to cell starvation and DNA-damaging stimuli compared to prokaryotes.

## 2.1 Molecular Evolution of Eukaryotic Beclin-1

To better understand the significance of Beclin-1 during eukaryotic evolution, we performed aggressive BLAST searches of genomic databases (see Supplementary file available on https://doi.org/10.1016/bs.acr.2017.11.002) using human Beclin-1 as a query sequence. The searches were performed locally using the approaches outlined in Rosenfeld and DeSalle (2012). Initially, human Beclin-1 was compared to the Beclin-1 peptide sequences of 10 critical taxa (Fig. 2A). These 10 critical taxa included *Arabidopsis* (plant), *Rhizopus* (fungi), *Nematostella* (Cnidaria), *Drosophila* (Protostome—Ecdysozoa), *Lottia* (Protostome—Lophotrochozoa), *Danio* (fish), *Xenopus* (amphibian), *Falco* (bird), *Felis* (cat), and *Homo* (human). Focusing upon the 1–133 5′ Beclin-1 amino acid sequences (Bcl-1), only mammalian and amphibian sequences were conserved while bird species (*Falco*) have lost Bcl-1. In the lower fungi, nonbilaterian animals, and protostomes (*Rhizopus*, *Nematostella*, *Drosophila*, and *Lottia*), 1–133 5′ Beclin-1 amino acid sequences displayed greater variability relative to mammalian Beclin-1. Much greater amino acid conservation was observed in the Beclin-1 Bcl-2 homology domain 3 (BH3; 114–130) amino acids as all animal taxa except for *Drosophila* contained the BH3 domain; BH3 was also not found in the plant taxa *Arabidopsis* (Fig. 2B). The evolutionary conservation of the Beclin-1 coiled-coil domain (CCD, 144–269 aa's), revealed a similar pattern to that of BH3; vertebrate taxa displaying similarity while the *Drosophila* and *Arabidopsis* taxa diverged from human Beclin-1 amino acid sequences (Fig. 2C). It is interesting to note that fungi, Cnidaria and the other protostome (*Lottia*) in this smaller dataset do not show the same high degree of divergence as *Arabidopsis* and *Drosophila* (Fig. 2D). In contrast to the BH3 and CCD domains, the ECD (244–337 aa's) was highly conserved between all studied taxa. While these 10 critical taxa were selected for their evolutionary relevance to mammalian evolution, additional sequence alignments were also performed from a richer diversity of 82 additional taxa for a total 92 studied taxa overall (Supplementary Fig. 1 in the online version at https://doi.org/10.1016/bs.acr.2017.11.002).

From these insights and deep sequencing data, an evolutionary tree of Beclin-1 proteins was constructed from plants, fungi, and animals (Fig. 3). This tree also shows where in the phylogeny interactors with Beclin-1 (PIK3C3, UVRAG, and miRNA-221) also arose. Starting from the earliest branch point, Beclin-1, ECD, and PIK3C3 appear to have arisen together during the first appearance of Beclin-1 in eukaryotes. Plant species do not have the Beclin-1 BH3 domain, while fungi do, which is interesting because Beclin-1 is not found in any known protist taxa. However, both plants and fungi do not display the Beclin-1 interacting protein ultraviolet radiation resistance-associated gene (UVRAG) which is interesting as plants and fungi are inundated with high doses of UV irradiation on average over their lifetimes. The Beclin-1 CCD domain, Bcl-1 domain, and miRNA-221, a negative regulator of Beclin-1, emerged in the common ancestor of vertebrates. The implications of these domains upon Beclin-1 function are summarized in Figure 4 (Fig. 4). The Bcl-1 domain was lost in the bird taxa later during vertebrate evolution (Fig. 3).

Beclin-1 was not found in prokaryotes (Bacteria, Archaea) as indicated in previous reports of Beclin-1 conservation in eukaryotic organisms (Wirawan et al., 2012). Interestingly however, Beclin-1 was also not found in protists like *Chlamydomonas* or algae, the common ancestors of plants, or in amoeba, *Plasmodium* or other ancestral protist lineages to fungi and animals. This occurs despite the fact that Beclin-1 is found in nearly all multicellular eukaryotes ranging from plants to fungi to sponges to cnidarians to bilateral animals (Supplementary Fig. 2 in the online version at https://doi.org/10.1016/bs.acr.2017.11.002). Furthermore, single celled eukaryotic organisms do not appear to require Beclin-1 expression, which makes the pattern of origin of Beclin-1 peculiar as multicellular eukaryotes (plants and animals) do not comprise a monophyletic group.

Considering these factors, there are at least three scenarios which could explain these patterns of Beclin-1 presence in the genomes of multicellular eukaryotes. (1) Beclin-1 may have been horizontally transferred from animals to plants or vice versa, which is very unlikely. (2) The common ancestor of all eukaryotes may have acquired Beclin-1. Subsequently, Beclin-1 may have been lost over time from these protist lineages which

**Fig. 4** Schematic of the major Beclin-1 protein domains and their respective protein–protein interaction partners. The *red–yellow shaded region* indicates the overlap between the Beclin-1 CCD and ECD domains. BH3, Bcl-2 homology-3 domain; CCD, coiled-coil domain; ECD, evolutionary conserved domain of Beclin-1.

may have been lost to the fossil record. This scenario would be especially appealing if Beclin-1 is deleterious to single celled life. Additional support for this argument was discussed by Huettenbrenner et al. in which single celled organisms were strongly selected against acquiring cell death effectors such as Beclin-1 at all costs while PCD appears to have only become advantageous in multicellular organisms such that individual cells could be sacrificed for the benefit of the organism as a whole (Huettenbrenner et al., 2003). In this scenario, we hypothesize that such selective pressures, including the maintenance of DNA genome fidelity in longer-lived multicellular eukaryotes, may have been sufficient to select for Beclin-1 expression. This scenario implies a hypothetical ancestral single celled protist having Beclin-1 in its genome and would be known as "the master ancestor" of all Beclin-1 containing multicellular eukaryotes or *dominus antecorris*. The downside of this scenario is that it would require several parallel losses of Beclin-1 in protists as these early protist lineages are very deep and not monophyletic. (3) Strong selective pressures may have driven all animals, and most multicellular plants such as angiosperms and clubmosses (Beclin-1 is absent in mosses and gymnosperms) to acquire Beclin-1. Prior to this analysis, the authors postulated that Beclin-1 arose as the first eukaryotic cell evolved. What we can hypothesize given the distribution of Beclin-1 in these eukaryotic genomes, is that natural selection has been a major driver of the evolution of this gene. Since the horizontal gene transfer scenario is highly unlikely, this leaves the most likely scenario of single origin with strong selection eliminating Beclin-1 in the genomes of protists, and the two-origin scenario with strong selection for convergence of the gene in plants and fungi/animals.

## 2.2 Beclin-1 and Eukaryotic Cell Autophagy

One of the principal functions of Beclin-1 is to facilitate the formation of the autophagosomes in response to cell stress (Fig. 1). Beclin-1 is regulated via phosphorylation at many sites including Ser-234, Ser-295, and Ser-90.

Beclin-1 phosphorylation at Ser-234 and Ser-295 inhibits Beclin-1 signaling, while its phosphorylation at Ser-90 activates Beclin-1 to induce autophagy (Fujiwara, Usui, Ohama, & Sato, 2016; Wang et al., 2012); Beclin-1 also appears to have three Tyr phosphorylation sites: 229, 233, and 352 associated with epidermal growth factor receptor (EGFR) while phosphorylation at Thr-388 is regulated by AMPK (Wei et al., 2013; Zhang et al., 2016). Furthermore, Beclin-1 is inhibited by complex formation with Bcl-2 (Klein et al., 2015). In this manner, Beclin-1 can be activated via the phosphorylation of Bcl-2 at the Ser-90 site, preventing the formation of the inhibitory Beclin-1/Bcl-2 complex via c-jun N-terminal kinase (JNK; Wei, Pattingre, Sinha, Bassik, & Levine, 2008; Zhou, Li, Jiang, & Zhou, 2015). AMP-activated protein kinase (AMPK) is one of the most well-known autophagy regulators; AMPK regulates autophagy by phosphorylating Beclin-1 at Thr-388. Thr-388 is required for autophagy to be induced during periods of cell starvation (Zhang et al., 2016). AMPK serves as an energy sensor to coordinate intracellular bioenergetic responses to cell starvation. When energy is low, AMPK activates Ulk1 via phosphorylation Ser-317 and Ser-777, inducing autophagy (Kim, Kundu, Viollet, & Guan, 2011). Beclin-1 is highly conserved with functional homology in yeast and higher-order eukaryotes including plants, reptiles, and humans, implicating its requirement for the growth and survival of eukaryotic cells as they complete their respective life cycles via the cooperation of these cell signaling pathways. In this review, the effects of Beclin-1 upon many aspects of cancer biology are discussed to illustrate the diverse roles of Beclin-1 to promote cancer therapeutic effects.

## 3. BECLIN-1 AND CANCER

Beclin-1 has been shown to be a haploinsufficient tumor suppressor in mice (Qu et al., 2003). Beclin-1 overexpression has also been shown in 85.2% of stage IIIB colon cancers (Li et al., 2009). Further analysis indicated that the overexpression of Beclin-1 in these colorectal tumors was associated with greater patient survival (Li et al., 2009). The controversial role of Beclin-1 upon cancer (summarized in Fig. 5) is primarily associated with its role as an autophagy inducer. Therefore, its influence upon oncogenesis and cancer therapy warrants further discussion of these phenomena. In this review, we will discuss the signaling pathways associated with Beclin-1 to determine how this complex tumor suppressive molecule affects oncogenesis, cancer prognosis, and cancer therapeutic responses clinically.

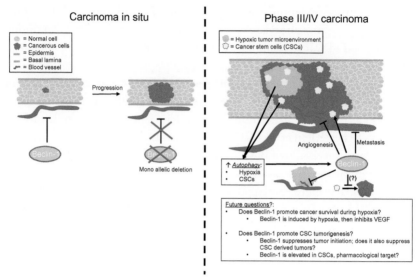

**Fig. 5** Schematic of the regulation of tumor growth, angiogenesis, and cancer stem cells by Beclin-1. Beclin-1 has been shown to inhibit tumor initiation in transgenic mice. Furthermore, Beclin-1 deletions are common in human cancers. When Beclin-1 is monoallelically deleted, tumor formation therefore becomes much more likely. In large (Phase III/IV) tumors, however, Beclin-1 and autophagy in general may play a role to enhance tumor malignancy. Interestingly however, Beclin-1 is associated with the inhibition of MMP-9 and VEGF to inhibit tumor angiogenesis and cancer cell metastasis despite the effects of tumor hypoxia to induce Beclin-1 and autophagy in vivo. Beclin-1 is also overexpressed in cancer stem cells (CSCs), however it remains unknown it Beclin-1 also inhibits tumor formation by CSCs as seen phenomenologically in primary tumors in transgenic mice. Perhaps Beclin-1 is an appealing target for the development of novel CSC targeted therapeutics; however, this will remain an area for further investigation.

## 3.1 Beclin-1 as a Tumor Suppressor

Beclin-1 has been implicated as a tumor suppressor, preventing tumor formation in many human tissues. However, the role of Beclin-1 as an autophagy inducer complicates this discussion as autophagy can be either tumor suppressive or oncogenic depending on the cellular context and genetic profile of each tumor (Bhutia et al., 2013; Liu & Debnath, 2016).

Clinically, Beclin-1 deletions are observed in approximately 40% of prostate, 50% of breast, and 75% of ovarian cancers clinically (Gao et al., 1995; Knudson, 1971; Qu et al., 2003). Decreased Beclin-1 expression is also frequently observed in human breast tumor tissue compared to patient matched normal controls (Liang et al., 1999). Furthermore, stable Beclin-1 expressing MCF-7 breast cancer cells were shown to display greater levels of autophagy, inhibiting colony formation in vitro, and decreased tumor growth in vivo (Gao et al., 1995). Beclin-1 heterozygous (+/−) mice were shown to display increased incidence of spontaneous tumor formation concurrent with decreased autophagy levels, establishing a clear role of autophagy, and Beclin-1 as a haploinsufficient tumor suppressor (Qu et al., 2003).

These phenomenological observations strongly implicate a tumor suppressive role of Beclin-1 in many differing tumor types. For the remainder of this review, when possible, Beclin-1 will be discussed within the context of breast cancer cells to focus upon Beclin-1 signal transduction. This is because Beclin-1 is a breast cancer tumor suppressor, its deletion enhances breast cancer malignancy (Liang et al., 1999), and Beclin-1 expression has been shown to be enhanced in breast CSC (Gong et al., 2013). Thus, breast cancer represents an excellent disease to understand the role of Beclin-1 upon malignancies in vitro and in vivo.

UVRAG contains four functional domains, a proline-rich domain, a lipid-binding domain, a C-terminal domain, and a Beclin-1-binding CCD (Gong et al., 2013; Zhao et al., 2012). UVRAG is an autophagy inducer and has been shown to be frequently deleted in breast, colorectal, and gastric cancers (Gong et al., 2013). Furthermore, the serine–threonine-specific protein kinase B (Akt) has been shown to inhibit UVRAG via mTOR (Yang et al., 2013); increased Akt/PI3K/mTOR signaling is thought to be oncogenic during breast tumorigenesis (Mohan et al., 2016). UVRAG is also inhibited by Rubicon (Kim et al., 2015), however, Rubicon has not been shown to promote tumorigenesis, to our knowledge, at this time. Considering these findings, UVRAG has been classified as a candidate haploinsufficient tumor suppressor in breast tissues (Zhao et al., 2012). Biochemical analysis of UVRAG-Beclin-1 binding indicates that UVRAG disrupts the inhibitory Bcl-2-Beclin-1 complex to induce autophagy (Noble, Dong, Manser, & Song, 2008). In this manner, the tumor suppressive effects of Beclin-1 may be suppressed indirectly by the deletion of UVRAG in Beclin-1 wild-type tumors of the breast. UVRAG also mediates the interaction of Beclin-1 with other downstream Beclin-1 signaling molecules to promote cancer therapeutic responses.

Bax-interacting factor-1 (Bif-1) is another tumor suppressor which has been shown to interact with Beclin-1 to inhibit breast cancer malignancy (Runkle, Meyerkord, Desai, Takahashi, & Wang, 2012). Bif-1 is also known as SH3GLB1 or Endophilin B1 (Cuddeback et al., 2001; Pierrat et al., 2001) and contains an N-terminal Bin–Amphiphysin–Rvs (BAR) domain and a C-terminal Src-homology 3 (SH3) domain. The Src oncogene has been shown to bind and phosphorylate Bif-1 on Tyr-80 to inhibit its cell signaling effects and the induction of apoptosis (Yamaguchi et al., 2008). The N-BAR domain is responsible for binding the plasma membrane, driving membrane curvature (Gallop et al., 2006; Masuda et al., 2006; Peter et al., 2004). Bif-1 has been shown to form a complex with Beclin-1 through UVRAG, regulating autophagosome formation (Takahashi et al., 2007). While the exact mechanism of Bif-1 and Beclin-1 interaction remains unknown, Bif-1 has been shown to suppress breast cancer cell migration by degrading EGFR (Runkle et al., 2012). Beclin-1 and EGFR have also been shown to directly interact with each other (Tan, Thapa, Sun, & Anderson, 2015). A survey of Bif-1 mutations in 284 cancerous tissue samples from various origins, revealed only one Bif-1 mutation (0.35%), indicating that Bif-1 mutations are rare in cancers (Kim, Yoo, & Lee, 2008). Interestingly however, Bif-1 expression has been shown to be downregulated in a subset of pancreatic adenocarcinomas (Coppola, Helm, Ghayouri, Malafa, & Wang, 2011) and in invasive and metastatic breast carcinoma in situ (Ho et al., 2009). Therefore, Bif-1 gene mutations may not contribute to tumorigenesis, however, it is likely that Bif-1 dysregulation may contribute to cancer progression via increased invasion and metastasis. Since Bif-1 mutations have not been observed in breast tumors, Bif-1 expression may also be silenced by gene methylation or epigenetic modifications in metastatic breast cancer cells. Even if such epigenetic changes in Bif-1 occurred it is unlikely such changes would establish a stable feedback-loop sufficient for malignant transformation, as shown by Hatziapostolou et al. in hepatocellular carcinoma (HCC; Hatziapostolou et al., 2011).

The activating molecule in Beclin-1-regulated autophagy protein 1 (Ambra1) is critical for autophagy induction and the regulation of Beclin-1 activity. Ambra1 also inhibits apoptosis driven cell death (Fimia, Corazzari, Antonioli, & Piacentini, 2013; Fimia et al., 2007; Gu et al., 2014). Ambra1 is expressed in the majority of pancreatic cancer patient tumors ($\sim$64%) and was significantly associated with poor overall survival ($P=0.032$; Ko et al., 2013). This finding was surprising as Ambra1 has been described as a haploinsufficient tumor suppressor in MEF's transformed with the RasV12/E1A oncogene. Furthermore, when mTOR was inhibited, Ambra1 enhanced the interaction of c-Myc and its phosphatase PP2A,

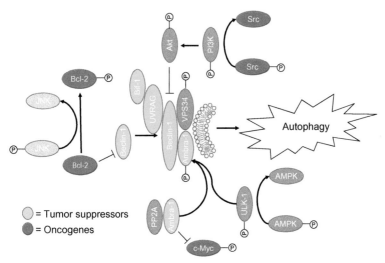

**Fig. 6** The regulation of Beclin-1 activity by a selection of oncogenes and tumor suppressive molecules. These molecules were *colored green* for tumor suppressors and *red* for oncogenes to indicate their interactions with Beclin-1 and autophagy. Molecules shown in *blue* are thought to be neutral in regard to oncogenesis.

inhibiting myc-induced oncogenesis (Cianfanelli et al., 2015). Consistent with these results, Ambra1 as an inducer of Beclin-1 and autophagy, is expected to stimulate cancer therapeutic responses via the Beclin-1 cell signaling axis alone or in combination with mTOR inhibitors such as the Rapalog family of drugs (Fig. 6).

Beclin-1 interacts with Ambra1 via its WD40 domain (Sun, 2016). In mice, the deficiency of Ambra1 results in significant neural tubular defects associated with autophagy impairment, accumulation of ubiquitinated proteins, increased apoptosis, and dysregulated cell division (Fimia et al., 2013). Conversely, the overexpression of Ambra1 in rapamycin-treated cells has been shown to significantly increase both rapamycin-induced and basal levels of autophagy in vitro (Fimia et al., 2013). Therefore, Ambra1 is clearly an autophagy inducer. Ambra1 is physiologically inactive when it associates with mTORC1. mTORC1 inhibits Ambra1 via phosphorylation at Ser-52, preventing the Ambra1 phosphorylation via ULK1 to promote Beclin-1-induced autophagy (Nazio et al., 2013). This study also demonstrated that the ULK1 and Beclin-1 complex crosstalk exists prior to autophagy induction, leading to autophagy-mediated cell signaling in response to cell starvation (Nazio et al., 2013). While the oncogenic effects of Ambra1 in pancreatic cancer requires further characterization, it is likely that Ambra1 is inactivated in pancreatic cancers which express high levels of activated KRAS and mTOR (Morran et al., 2014), which may account for these clinical observations. Additionally, Ambra1 is predominantly expressed in neural tissue so the effects of Ambra1 expression in glioma, and nonneurological malignances, remains an intriguing area for further study.

These tumor suppressive molecules, Beclin-1, UVRAG, Bif-1, and Ambra1, induce autophagy via the direct or in-direct stimulation of Beclin-1 to mediate their cancer therapeutic effects. As described with Ambra1, these molecules can be incriminated as tumor suppressors or oncogenic factors depending on the tissue or the activation of diverse cell signaling contexts, respectively. Before addressing the effects of these and other molecules on enhancing cancer malignancy, we will speak globally about these proteins and their interactions to induce cytotoxic autophagy-dependent cancer cell death. When Bcl-2 is phosphorylated and mTOR is inhibited, Beclin-1 forms an autophagy inducing complex, which together with UVRAG, Bif-1, and Ambra1, inhibits the activity of the oncogenes Bcl-2, Src, and myc to repress tumor malignancy and kill cancer cells (Fig. 6). Therapeutically, stimulating this pathway has been effective using the drug Bergapten in breast cancer cells (De Amicis et al., 2015). Bergapten is a natural psoralen derivative present in many fruits and vegetables. In this study, Bergapten was shown to induce cytotoxic autophagy via the upregulation of phosphatase and tensin homolog (PTEN), p38MAPK/NF-Y, Beclin-1, PI3KIII, UVRAG, and Ambra1, and downregulate mTOR signaling (De Amicis et al., 2015). PTEN is a tumor suppressive molecule which induces autophagy and arrests cell cycle progression (Arico et al., 2001), while suppressing the oncogenes PI3K and AKT, respectively (Arico et al., 2001; Petiot, Ogier-Denis, Blommaart, Meijer, & Codogno, 2000). For cancers with dysregulated mTOR and Bcl-2 expression, targeting the Beclin-1 pathway appears to induce profound cancer therapeutic effects for the treatment of cancer clinically.

## 3.2 The *mda-7*-miR-221-Beclin-1 Axis

One of the genes known to have antitumor activity in a broad range of malignancies is *mda-7/IL-24* (Cunningham et al., 2005; Fisher, 2005; Fisher et al., 2003). This IL-10 gene family member (Dash et al., 2010) was discovered as a tumor suppressor using subtraction hybridization in terminally differentiated human melanoma cells (Jiang & Fisher, 1993; Jiang,

Lin, Su, Goldstein, & Fisher, 1995). Overexpression of *mda-7/IL-24* has been shown to sensitize cancer cells to conventional chemotherapeutic drugs (Chada et al., 2006; Emdad, Lebedeva, Su, Gupta, et al., 2007; Emdad, Lebedeva, Su, Sarkar, et al., 2007) and radiotherapeutics (Nishikawa, Ramesh, Munshi, Chada, & Meyn, 2004; Su et al., 2003; Yacoub et al., 2003). One of the most attractive features of *mda-7/IL-24* is its capacity to induce cancer-selective cell death, as well as its role as an immune-stimulatory molecule killing distant metastasis via bystander effects (Gao et al., 2008; Sarkar et al., 2002; Sauane et al., 2008; Su et al., 2005, 1998). Furthermore, *mda-7/IL-24* also blocks angiogenesis (Nishikawa et al., 2004; Ramesh et al., 2003) slowing tumor growth and decreasing the risk of cancer metastasis. The cancer-specific cell death effects of *mda-7/IL-24* have been shown to be mediated through its interaction with the chaperone protein BiP/GRP78, initiating UPR or unfolded protein response (Gupta et al., 2006). Some of the other signaling pathways and molecules downstream of *mda-7/IL-24* which have been described in the literature include AIF (Bhoopathi et al., 2016), miR-221 (Pradhan et al., 2017), and Beclin-1 (Bhutia et al., 2010; Pradhan et al., 2017). Owing to its in vitro and in vivo anticancer activity, *mda-7/IL-24* has been successfully translated into early phase cancer clinical trials (Cunningham et al., 2005; Emdad et al., 2009; Fisher et al., 2003, 2007; Lebedeva et al., 2005; Tong et al., 2005).

MicroRNAs are noncoding small, 19–22 nucleotide RNA molecules which are key regulators of gene expression. MicroRNAs negatively regulate gene expression by selectively binding and degrading target messenger RNA (mRNA) or by inhibiting mRNA translation. Recent studies have shown that microRNAs are also regulated by *mda-7/IL-24*. In a human melanoma model, it was shown that *mda7/IL-24* downregulates miR-221 (Das et al., 2010). Recent studies from our group also indicated that miR-221 transcriptionally targets Beclin-1 (Pradhan et al., 2017). Beclin-1 is also transcriptionally regulated by a number of other cellular pathways. p65, a NF-κB pathway family member, induces Beclin-1 (Copetti, Demarchi, & Schneider, 2009). 14-3-3 protein and E2F1 also transactivate Beclin-1 leading to the upregulation of autophagy (Wang, Ling, & Lin, 2010). Previous studies have also shown the role of microRNAs in Beclin-1 regulation. miR-30a targets Beclin-1 UTR and regulates the autophagy process (Zhu et al., 2009). Overexpression of miR-30a downregulates Beclin-1 transcripts as well as protein levels. MicroRNA-221 is one of the most frequently upregulated miRs in cancer. It targets a number of tumor

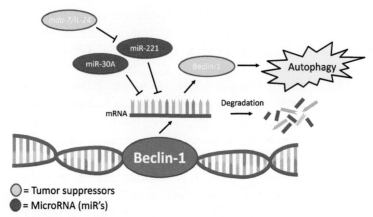

**Fig. 7** The regulation of Beclin-1 expression via microRNAs and the effect of mda-7/IL-24 to increase the activity of Beclin-1 to induce autophagy and lyse cancer cells. *Green circles* indicate tumor suppressive molecules. *Red circles* are indicative of microRNA. The *blue double-stranded molecule* is a cartoon of the Beclin-1 gene. The single-stranded molecule with the *red* backbone is indicative of the Beclin-1 mRNA transcript. The debris is indicative of mRNA degradation following the association of miRs with Beclin-1 mRNA transcripts.

suppressors including p27 (le Sage et al., 2007), PUMA (Zhang et al., 2010), PTEN (Garofalo et al., 2009), and p57 (Fornari et al., 2008). Beclin-1 can also be transcriptionally targeted by miR-221 (Pradhan et al., 2017), blocking toxic autophagic cell death. Overall this regulatory loop identified a new pathway of *mda-7/IL-24*-mediated cancer-specific toxic autophagy leading to cell death (Fig. 7).

### 3.3 Arguments for and Against the Oncogenicity of Beclin-1

Enhanced autophagy signaling via Beclin-1 may also promote cancer malignancy during periods of cell stress imposed by hypoxia and the metastatic process. The phenomenological observations from breast cancer patients indicate that Beclin-1 expression declines with increasing cancer stages, however, some reports suggest Beclin-1 and autophagy may contribute to resistance to cancer chemotherapy (Ying et al., 2015) and radiation (Apel, Herr, Schwarz, Rodemann, & Mayer, 2008). Beclin-1 may play a role in breast CSC maintenance which are refractory to chemotherapy and radiation treatment (Gong et al., 2013). These possibilities will be discussed and the contribution of Beclin-1 in promoting cancer oncogenesis and tumor malignancy in greater detail in this review.

### 3.3.1 Beclin-1 Expression and Cancer

Beclin-1 monoallelic deletions have been observed in many tumors implicating Beclin-1 as a tumor suppressive molecule. A more detailed description of Beclin-1 in cancer is shown in Table 1 (Ahn et al., 2007; Cliby et al., 1993; Ding et al., 2008; Eccles et al., 1992; Futreal et al., 1992; Gao et al., 1995; Koukourakis et al., 2010; Li et al., 2009; Li, Chen, et al., 2010; Lin et al., 2013; Miracco et al., 2010, 2007; Pirtoli et al., 2009; Radwan et al., 2016; Russell et al., 1990; Saito et al., 1993; Tangir et al., 1996; Zhang et al., 2009; Zhou et al., 2013). While most studies indicate that Beclin-1 is tumor suppressive, it is possible that the deletion of Beclin-1 may have led to the loss of autophagy signaling, and thereby stimulated many other pathways including necrosis and inflammation which are associated with tumorigenesis (Mathew & White, 2011; White & DiPaola, 2009). Interestingly, Beclin-1 biallelic deletions have not been observed in tumors clinically, failing the Knudson two-hit hypothesis required for genes to be characterized as true tumor suppressors (Knudson, 1971). Therefore, Beclin-1 is described as a haploinsufficient tumor suppressor instead as Beclin-1 heterozygous mice displayed greater tumor formation (Qu et al., 2003). The lack of biallelic Beclin-1 deletions in tumors clinically also indicates that expression of at least one functional Beclin-1 allele is required for cancer cell survival, or at least, for tumor growth.

Qu et al. rigorously tested whether or not Beclin-1 functions as a tumor suppressor or an oncogene by crossing mice which express the oncogenic hepatitis B virus (HBV) large-envelope polypeptide with Beclin-1 wild-type or heterozygous mice, respectively (Qu et al., 2003). These data indicated that 15% of Beclin-1 heterozygous mice developed palpable tumors with confirmed histological malignancy vs only 1% of Beclin-1 wild-type littermates that formed tumors at 13–18 weeks of age. One hundred mice were examined in both the Beclin-1 wild-type and heterozygous groups, respectively. These malignancies were consistent with lymphoma, lung, and liver cancers. Beclin-1 heterozygote mice also displayed tissue hyperplasia in the mammary gland and in splenic germinal centers. Beclin-1 heterozygous tumors also displayed greater overall cell division and larger tumor formation (Qu et al., 2003). Finally, Beclin-1 heterozygous mice were shown to display less autophagy activity in bronchial epithelial cells and germinal B lymphocytes. Tumors were later observed to form from these tissues (bronchus and lymphocytes), indicating that autophagy induced by Beclin-1 was protective against lung and lymphoma oncogenesis in mice expressing the HBV oncogene (Qu et al., 2003). This study was carefully designed and

Table 1 The Expression of Beclin-1 in Cancers Clinically

| Cancer Types | Dysregulation of Beclin-1 Expression | Clinical Correlations | References |
|---|---|---|---|
| Acute Lymphocytic Leukemia (ALL) | Decreased Beclin-1 expression | Beclin-1 extended ALL patient survival seemingly by inhibiting HIF-1α | Radwan, Hamdy, Hegab, and El-Mesallamy (2016) |
| Brain Cancer | Expression of Beclin-1 mRNA is decreased in higher-grade tumors | Low Beclin-1 correlates with poor survival | Lin et al. (2013) and Miracco et al. (2007) |
| Brain Cancer | Cytoplasmic Beclin-1 was positively associated with apoptosis and the inhibition of cell proliferation | High cytoplasmic Beclin-1 expression is associated with better survival and high Karnofski classification values | Pirtoli et al. (2009) |
| Breast Cancer | The locus which harbors Beclin-1 is monoallelically deleted in ~50% of cases | ND | Futreal et al. (1992) and Saito et al. (1993) |
| Breast Cancer | Decreased Beclin-1 mRNA and protein expression, monoallellic deletions, promoter methylation, and occurred in breast cancer tissues | Higher-Beclin-1 mRNA expression in BRCA1-positive tumors | Li, Chen, et al. (2010) |
| Colorectal Cancer | Underexpression and extensive overexpression of Beclin-1 were observed in 15.5% and 21.3% of cases, respectively | Dysregulated Beclin-1 was associated with poorer survival; "normal-like" Beclin-1 expression improved survival | Koukourakis et al. (2010) |
| Colorectal Cancer | Beclin-1 expression was greater in cancerous than matched normal tissue in 95% of cases | ND | Ahn et al. (2007) |
| Colorectal Cancer | Strong expression of Beclin-1 was detected in stage IIIB tumors but not adjacent normal tissues | Higher-Beclin-1 expression is associated with longer survival | Li et al. (2009) |

| | | | |
|---|---|---|---|
| Gastric Cancer | Beclin-1 expression was greater in cancerous than matched normal tissue in 83% of cases | ND | Ahn et al. (2007) |
| Liver Cancer | A >twofold decrease in Beclin-1 mRNA was observed in 45.5% HCC tissues; decreased Beclin-1 protein expression was confirmed in six out of eight cases | Lower levels of Beclin-1 mRNA was correlated with recurrence. Decreased patient survival was observed in Beclin-1 negative, Bcl-xL–positive tumors | Ding et al. (2008) |
| Lung Cancer | Beclin-1 was significantly lower in NSCLC compared with adjacent normal controls | Decreased Beclin-1 lead to increased recurrence, decreased overall survival | Zhou et al. (2013) |
| Melanoma | Beclin-1 and LC3 are downregulated during disease progression | Beclin-1 and LC3 mRNA levels can discriminate between nonmalignant and malignant lesions | Miracco et al. (2010) |
| Osteosarcoma | Beclin-1 protein levels are lower (by IHC) in osteosarcoma than in normal bone | ND | Zhang et al. (2009) |
| Ovarian cancer | The locus harboring Beclin-1 is monoallelically deleted in ~75% of cases | ND | Cliby et al. (1993), Eccles et al. (1992), Russell, Hickey, Lowry, White, and Atkinson (1990), and Tangir et al. (1996) |
| Prostate cancer | The locus harboring Beclin-1 is monoallelically deleted in ~40% of cases | ND | Gao et al. (1995) |

Beclin-1 plays a significant role in the regulation of tumor growth in numerous malignancies. ND indicates not determined.

categorically disproved the hypothesis that Beclin-1 promotes tumor oncogenesis, indicating that the downregulation of Beclin-1 expression both promoted tumor cell growth and enhanced tumor formation in vivo. Consistent with these observations, Beclin-1 may also serve as a tumor suppressor gene due to its ability to control the cell cycle (Sun et al., 2011) and stimulate apoptosis (Li et al., 2013).

Alternatively, when Beclin-1 is suppressed and autophagy is therefore dysregulated, tumorigenesis may occur directly via other autophagy-related molecules or indirectly via the formation of oxidative free radicals. Although the crosstalk between mitochondrial dysfunction, autophagy, and redox signaling are not well understood (Lee, Giordano, & Zhang, 2012), these concepts will be discussed further here. For example, p62/SQSTM1, which binds ubiquitinated proteins and sequesters them into the autophagolysosome for degradation, has been shown to be upregulated along with other autophagy genes by nuclear factor-erythroid 2-related factor 2 (Nrf2) nuclear translocation and redox signaling (Giles, Gutowski, Giles, & Jacob, 2003; Riley et al., 2010). By this mechanism, Nrf2 has been shown to increase p62 expression, leading to p62-mediated reactive oxygen species scavenger activity in models of neurodegeneration (Lee et al., 2012).

Relating these concepts back to cancer, it is hypothesized that deficiencies in autophagy which could lead to the accumulation of oxidative radicals, contributing to overall genome instability and tumorigenesis. Interestingly however, the upregulation of p62 was only shown to be oncogenic when autophagy is suppressed (Mathew et al., 2009). Furthermore, p62 is essential for HER2-driven oncogenic breast transformation via multiple signaling pathways, including the PTEN/PI3KIII/AKT axis, WNT/β-catenin, NF-κB, and NRF2-KEAP1 pathways, respectively (Cai-McRae, Zhong, & Karantza, 2015). For example, Beclin-1 has been shown to interact with Her2 and when this interaction is disrupted by the dual tyrosine kinase inhibitor, lapatinib, autophagy is induced to lyse Her2 expressing breast cancer cells via the phosphorylation of Akt (Han et al., 2013). These studies indicate that the disruption of p62-mediated oxidative scavenging is cancer therapeutic and suppresses the formation of breast tumors. The cancer therapeutic consequences of autophagy disruption via Beclin-1 and p62 deletion, suggests a dual role of autophagy during carcinogenesis such that autophagy suppresses tumor initiation (Kung, Budina, Balaburski, Bergenstock, & Murphy, 2011), but autophagy also may support the maintenance of established tumors and their respective microenvironments during hypoxia (Tan et al., 2016), further enhancing tumor invasion and metastasis (Mowers, Sharifi, & Macleod, 2017); these concepts are summarized as illustrated in Fig. 5.

## 3.4 Beclin-1 Expression and Its Effect Upon Cancer Progression

Overall, the expression of Beclin-1 is not clearly implicated as repressed or overexpressed in colorectal cancers clinically (Koukourakis et al., 2010). In T-cell lymphoma, however, decreased Beclin-1 expression associates with decreased overall and progression-free survival indicating tissue-specific consequences of Beclin-1 upon cancer malignancy (Huang et al., 2010). Therefore, to elucidate the effect of Beclin-1 upon colorectal tumors clinically, the effect of Beclin-1 expression upon colorectal adenocarcinoma prognosis was assessed in 155 patients treated with surgery alone (Koukourakis et al., 2010). Beclin-1 expression was shown to be at normal levels ($n=62$, 40%), under expressed ($n=24$, 15.5%), limited overexpression ($n=36$, 23.2%), and extensively overexpressed ($n=33$, 21.3%) in these 155 colorectal adenocarcinoma patients by tumor histopathological analysis. Patient 3-year overall survival was the greatest in the normal-like Beclin-1 expression group (89%) compared to Beclin-1 limited overexpression (77.7%), Beclin-1 extensive overexpression (43.3%), however, the lowest patient survival was shown in the Beclin-1 under expression group (31%). The authors indicated that two biologically distinct pathways of Beclin-1 activity, both linked with tumor aggressiveness and that the maintenance of physiological levels of Beclin-1 in cancers is relevant in colon cancer (Koukourakis et al., 2010). First, the loss of Beclin-1, by monoallelic gene deletion, was shown to stimulate breast oncogenesis (Aita et al., 1999). Second, as Beclin-1 interacts with the antiapoptotic Bcl-2 protein (Cao & Klionsky, 2007), the loss of Beclin-1 may have potentiated the antiapoptotic machinery which may account, at least in part, for the poorer prognosis of Beclin-1 negative colorectal tumors. Furthermore, Li et al. found that increased Beclin-1 expression is linked with better stage III colorectal cancer patient prognosis (Liang, Yu, Brown, & Levine, 2001). Beclin-1 was also shown to be a crucial regulator of colorectal cancer growth and metastasis upon further study (Koukourakis et al., 2010). Therefore, additional insights into the role of Beclin-1 to promote cancer malignancy are critical to understanding this process mechanistically.

### 3.4.1 Beclin-1 in Cancer Invasion, Metastasis, and Tumor Hypoxia

Autophagy is known to support the survival of cancer cells in hypoxic tumor microenvironments (Mathew & White, 2011). However, Beclin-1 has also been shown to inhibit the proliferation, invasion, and metastasis of CaSki cervical cancer cells in vivo (Sun et al., 2011). In this cancer model system, Beclin-1 expression was shown to arrest cancer cells in the G0/G1 phase of the cell cycle and to inhibit vascular endothelial growth factor (VEGF) and matrix metalloprotease 9 (MMP-9), which likely was responsible for the observed repression of invasion and metastasis in CaSki tumors which stably overexpress Beclin-1 relative to pcDNA3.1 (Sun et al., 2011). Beclin-1 overexpression was also shown to suppress angiogenesis, VEGF, and MMP-9 expression in vitro (Sun et al., 2011). Interestingly, however, in Beclin-1 heterozygous mice, differences in circulating VEGF were not observed whereas the production of EPO was markedly elevated in the circulation of Beclin-1 heterozygotes relative to Beclin-1 wild-type mice (Lee, Kim, Jin, Choi, & Ryter, 2011). It is possible therefore that Beclin-1 only suppresses VEGF in vitro or in the tumor microenvironment. Taken together, these data suggest that Beclin-1 plays a role in preventing cancer malignancy via the suppression of cancer cell growth, invasion, and metastasis, while simultaneously preventing tumor angiogenesis.

Consistent with these findings, heterozygous disruption of Beclin-1 accelerated tumor growth and angiogenesis under hypoxic conditions in melanoma bearing mice in vivo (Lee et al., 2011). Cells cultured from Beclin-1 wild-type and Beclin-1 heterozygous mice indicated that Beclin-1 expression inhibited angiogenesis via decreased hypoxia-inducible factor-2α (HIF-2α) expression, limiting endothelial cell proliferation, and tube formation in response to hypoxia (Lee et al., 2011). Furthermore, these data demonstrate that mice deficient in Beclin-1 displayed a proangiogenic phenotype associated with HIF-2α upregulation and increased erythropoietin production (Lee et al., 2011). Conversely, however, when colon cancer patient samples were analyzed histologically, these data suggested high levels of Beclin-1 expression was associated with improved survival (Li et al., 2009), while an in depth metaanalysis of data from six clinical studies indicated that elevated Beclin-1 expression was associated with tumor metastasis and poor prognosis in colorectal cancer patients (Han et al., 2014). On a molecular level, Beclin-1 negatively regulates angiogenesis in vivo in murine model systems, with implications for the inhibition of tumor growth (Sun et al., 2011). While the role of Beclin-1 in colorectal cancer pathology remains controversial, it is clear that Beclin-1 is an intriguing molecular target for colorectal and many other human malignancies. These competing findings may also reflect the differential effects of Beclin-1 depending on its subcellular compartmentalization in these individual tumor cells or the expression of other Beclin-1 regulatory molecules.

## 3.5 Beclin-1 and Resistance to Cancer Therapy
### 3.5.1 The Effect of Beclin-1 Upon Cancer Chemotherapeutic Efficacy and Patient Survival

While Beclin-1 possesses numerous tumor suppressive mechanisms of action, the effects of Beclin-1 on autophagy induction may antagonize the efficacy of cytotoxic cancer therapeutics acting through apoptosis. These antagonistic effects of autophagy upon apoptosis are likely a consequence of numerous individual molecular interactions (Moreau, Luo, & Rubinsztein, 2010). Two examples are the mutually inhibitory interactions between LC3 and Beclin-1 with the apoptosis inducing and effector caspase, caspase-3 (Ma, Zhang, Huang, Guo, & Hu, 2016). Interestingly, while autophagy has been implicated in enhancing cancer cell survival via the inhibition of apoptosis (Moreau et al., 2010), such antagonism has not be observed via the autophagy inducer Beclin-1; as determined via selectively silencing Beclin-1 expression with siRNA (Huang et al., 2014). In this manner, the role of Beclin-1 upon cancer therapeutic responses are predicated upon many factors, reflecting the Janus faces of autophagy and its effects upon Beclin-1 signaling.

The effects of Beclin-1 expression upon cancer therapeutic responses of colorectal and ovarian cancer patients treated with the apoptosis inducers cetuximab or carboplatin and paxilitaxel, respectively, were determined in the clinical setting. Colorectal cancer patients treated with cetuximab-containing or noncetuximab-containing chemotherapy were then evaluated for their respective cancer therapeutic responses and Beclin-1 expression in treated tumors (Guo et al., 2011). Cetuximab is a monoclonal antibody which inhibits the EGFR and has been approved for use in many human cancers. Cetuximab has been reported to induce autophagy via the suppression of the mammalian target of rapamycin (mTOR) (Li, Lu, Pan, & Fan, 2010), however, cetuximab has also been implicated as an apoptosis inducer via the inhibition of the mitogen-activated protein kinase (MAPK) and the Janus kinase/STAT3 pathways in head and neck squamous cell carcinoma cells (Bonner et al., 2000). In colorectal cancer patients treated with cetuximab, decreased Beclin-1 expression was associated with prolonged progression-free survival (Guo et al., 2011). The authors indicated that the role of autophagy in cetuximab treatment is controversial and that increased or decreased levels of Beclin-1 expression may not be indicative of the presence or absence of autophagy in treated tumors (Guo et al., 2011). However, decreased LC3 expression was correlated with greater overall response rates for cetuximab-treated colorectal cancer patients (Guo et al., 2011). In ovarian cancer patients treated with surgical resection followed by treatment with the cytotoxic cancer chemotherapeutics carboplatin and paclitaxel; patient cancer therapeutic responses and Beclin-1 expression in treated tumors were also evaluated (Valente et al., 2014). Carboplatin is a platinum-based chemotherapeutic and cell cycle nonselective inducer of DNA damage and apoptosis. Paclitaxel is a cell cycle selective inhibitor of cell division which stabilizes microtubules by binding β-tubulin polymers, preventing cell division, and inducing apoptosis via prolonged activation of the mitotic cell cycle checkpoint. No differences were observed at 5 years' posttreatment, however, a greater proportion of ovarian cancer patients whose tumors expressed higher levels of Beclin-1 displayed significantly greater survival ($n=34$, 83%; $P$-value $< 0.03$) relative to patients with tumors expressing less Beclin-1 ($n=7$, 17%; Valente et al., 2014). Therefore, Beclin-1 appears to antagonize the cancer therapeutic efficacy of cetuximab in colorectal cancer cells while Beclin-1 expression appears to enhance the cancer therapeutic effects of the apoptosis inducers carboplatin and paclitaxel in ovarian cancer patients. Similarly, Beclin-1 overexpression enhanced the sensitivity of CaSki cervical cancer cells to apoptosis via cisplatin, paclitaxel, 5-FU, and epirubicin (Sun et al., 2010). Unfortunately, to the best of our knowledge, no studies have compared the same cancer chemotherapeutics upon the survival of cancer patients with differing tumor types investigating the effects of Beclin-1 as a clinical biomarker of overall cancer patient survival, indicating the need for further research in this area. Therefore, it cannot be concluded at this time if either fundamental differences exist between Beclin-1 expression and the oncogenesis of ovarian and colorectal cancers or if yet undiscovered molecular markers for response to these drugs may explain the differences observed between these cancer therapeutics, respectively.

### 3.5.2 The Effect of Beclin-1 Upon Radiation Therapy, Autophagy, and UVRAG

Growing evidence suggests that autophagy contributes to the resistance of cancers to radiation therapy (Gewirtz, 2014; Wilson et al., 2011; Yin et al., 2011), however, the effects of the Beclin-1 pathway upon radiation-resistance remains undefined. Beclin-1 has been shown to improve DNA stability by physical interactions with UVRAG (Park, Tougeron, Huang, Okamoto, & Sinicrope, 2014). Very recently, Beclin-1-mediated DNA stability was shown to occur independently of autophagy via interaction with UVRAG to enhance cellular DNA damage repair and maintain genome integrity (Xu et al., 2017; Zhao et al., 2012). Likewise, increased UVRAG expression has been shown to protect cancer cells against radiation treatment (Park et al., 2014). UVRAG also disrupts irradiation-induced apoptosis via altering Bax localization thereby inhibiting apoptosis (Yin et al., 2011), when Beclin-1 translocates to the nucleus. Consequently, the downregulation of UVRAG expression inhibits autophagy and sensitizes cancer cells to DNA damage-mediated cell death via the upregulation of serine–threonine-specific protein kinase B (Akt) and mTOR (Yang et al., 2013).

The clinical application of radiation therapy entails the targeted exposure of tumors to high doses to cytotoxic doses of ionizing radiation, leading to the formation of oxidative radicals and DNA damage and eventually apoptotic cell death via p53 (Cui et al., 2016). In K-ras-induced transgenic murine lung tumors, however, treatment with Beclin-1 aerosols was shown to sensitize these tumors to fractionated radiation treatment (Shin et al., 2012). This result was surprising considering the role of the autophagy proteins Beclin-1, Atg5, and Atg12 to protect against DNA damage (He, Dai, Jin, Liu, & Rent, 2012). However, if the effects of Beclin-1 and UVRAG upon autophagy are independent from its effects upon radiation-induced cell death, as recent reports suggest (Xu et al., 2017; Zhao et al., 2012), additional molecules may predict the consequences of Beclin-1 and UVRAG signaling in cancerous tissues.

Returning to the discussion of the K-ras-induced lung carcinogenic model (Shin et al., 2012), by supplementing these cells with Beclin-1 containing aerosols, the resistance of these tumors to radiation treatment inherent to K-ras expressing tumors was overcome (Bernhard et al., 2000). It is likely that Beclin-1 stimulated the death of tumors following radiation treatment by stimulating DNA damage responses in vivo. Phenomenologically, radiation treatment does not appear to depend upon autophagy. For example, while autophagy inhibition rendered some breast cancer cells such as MCF-7 and ZR-75 more sensitive to radiation (Bristol et al., 2012; Wilson et al., 2011), the inhibition of autophagy in 4H1 and HS578t breast cancer cells had no effect upon their viability posttreatment (Bristol et al., 2013; Gewirtz, 2014). Mechanistically, the inhibition of autophagy has only been shown to enhance radiation efficacy in cancer cells with functional P53 (Chakradeo et al., 2015). Beclin-1 and UVRAG have also been shown to protect cells against radiation treatment (Park et al., 2014), however, the antagonism of autophagy may only be therapeutic due to the inhibitory crosstalk of the downstream apoptosis autophagy-signaling molecules of the respective DNA damage response genes P53 and Beclin-1, respectively (Ma et al., 2016). Therefore, it appears that Beclin-1 expression supports the lysis of cancer cells in response to DNA damage, however, with careful consideration the expression profiles of p53 and UVRAG, the development of more personalized radiation treatment protocols is feasible to improve patient clinical outcomes from radiation treatment.

## 3.6 Beclin-1 and CSC Maintenance

One fundamental problem with cancer chemotherapy and radiation treatments is the inherent resistance of CSCs to these therapeutic approaches. Gong et al. created three-dimensional cultures of breast cancer cells called "mammospheres" to study the biology of breast CSC containing tumors in an easily manipulated model system in vitro. Using this approach, Beclin-1 expression and autophagic flux were enhanced in cancer cells grown in spherical vs monolayer cultures (Gong et al., 2013). Beclin-1 expression was enhanced further in mammosphere cultures in the presence of CSCs (Gong et al., 2013). Using models of Beclin-1 suppression (tet-off, shRNA), decreased Beclin-1 expression was shown to inhibit the formation of spherical cultures by cancer cells and CSCs in vitro; decreased Beclin-1 also decreased tumor volumes produced from implanted mammospheres in vivo (Gong et al., 2013). Taken together, these data indicated that Beclin-1 expression was required for breast CSC-induced tumorigenesis.

Considering that enhanced Beclin-1 expression was associated with spherical cancer cell growth and was upregulated in the presence of CSCs, additional insight is necessary to elucidate the cancer therapeutic or oncogenic effects of Beclin-1 expression upon CSCs. One of the best validated markers for stem/progenitor-like cells is the expression of aldehyde dehydrogenase 1 (ALDH1; Martin et al., 2016). High levels of ALDH1 expression can correctly identify tumorigenic cell populations prior to implantation and is correlated with poor overall prognosis in cancer patients (Martin et al., 2016). At this time, Beclin-1 has not been shown to interact with ALDH1 or other cancer stem cell markers mechanistically. However, the close association of enhanced Beclin-1 expression in ALDH1-positive cells indicates that a potential direct or indirect signaling axis may exist between these two molecules. Therefore, Beclin-1 appears to contribute to the maintenance of CSCs and may serve as an intriguing molecular target for their destruction.

## 4. CONCLUSIONS AND FUTURE DIRECTIONS

The tumor suppressive properties of Beclin-1 appear to be a result of complex series of interactions with autophagy and apoptotic cell death pathways in cooperation with other tumor suppressive molecules such as UVRAG, Bif-1, Ambra1, and *mda-7/IL-24*. Evolutionarily, Beclin-1 appears to have emerged as multicellular eukaryotes needed additional biochemical pathways by which to adapt to cell stress, maintain the integrity of their DNA genomes, or provide PCD mechanism to enhance the survival of multicellular organisms. In this review, we have also postulated the evolutionary events in the molecular evolution of Beclin-1. The distribution of Beclin-1 in eukaryotes requires parallel losses in protest lineages or parallel gains of the gene in plant and fungi/animal genomes. Which scenario is correct will require further investigation. In human cancers, Beclin-1 is frequently monoallelically deleted and, it is implied that, one functional copy is always present in human cells and this remaining allele must not be silenced via methylation or epigenetic modifications for the cancer cells to survive and grow. Furthermore, the deletion of both Beclin-1 alleles has not been observed in tumors, clinically supporting this claim. The overexpression of Beclin-1 has also been implicated in inhibiting angiogenesis and tumor metastasis. One of the most prominent and growing concerns in modern oncology is the elimination of CSC. Beclin-1 expression has been shown to be elevated in these cells, comprising an intriguing molecular target for the development of novel CSC treatment modalities. Although clearly complex, defining the precise role of Beclin-1 in mediating cancer cell development, response to therapy and progression will be pivotal in determining if manipulating this molecule can be used to enhance therapeutic outcomes.

## ACKNOWLEDGMENTS

Support for our laboratories was provided in part by National Institutes of Health Grants R01 CA097318 (P.B.F.), R01 CA168517 (Maurizio Pellecchia and P.B.F.), GM093857 VCU IRACDA (P.B.F. and Joyce A. Lloyd), and P50 CA058326 (M.G.P. and P.B.F.); the Samuel Waxman Cancer Research Foundation (P.B.F. and D.S.); National Foundation for Cancer Research (P.B.F.); NCI Cancer Center Support Grant to VCU Massey Cancer Center P30 CA016059 (P.B.F.); and VCU Massey Cancer Center developmental funds (P.B. F.). P.B.F. and D.S. are SWCRF investigators. P.B.F. holds the Thelma Newmeyer Corman Chair in Cancer Research in the VCU Massey Cancer Center. R.D. thanks the Sackler Institute for Comparative Genomics, the Simons Foundation, and the Korein Foundation for continued support.

*Conflict of interest*: Dr. Fisher is a cofounder and has ownership interest in Cancer Targeting Systems (CTS), Inc.

## SUPPLEMENTARY MATERIAL

Supplementary data to this article can be found online at https://doi.org/10.1016/bs.acr.2017.11.002.

## REFERENCES

Ahn, C. H., Jeong, E. G., Lee, J. W., Kim, M. S., Kim, S. H., Kim, S. S., et al. (2007). Expression of beclin-1, an autophagy-related protein, in gastric and colorectal cancers. *APMIS*, *115*(12), 1344–1349. https://doi.org/10.1111/j.1600-0463.2007.00858.x.

Aita, V. M., Liang, X. H., Murty, V. V., Pincus, D. L., Yu, W., Cayanis, E., et al. (1999). Cloning and genomic organization of beclin 1, a candidate tumor suppressor gene on chromosome 17q21. *Genomics*, *59*(1), 59–65. https://doi.org/10.1006/geno.1999.5851.

Apel, A., Herr, I., Schwarz, H., Rodemann, H. P., & Mayer, A. (2008). Blocked autophagy sensitizes resistant carcinoma cells to radiation therapy. *Cancer Research*, *68*(5), 1485–1494. https://doi.org/10.1158/0008-5472.CAN-07-0562.

Arico, S., Petiot, A., Bauvy, C., Dubbelhuis, P. F., Meijer, A. J., Codogno, P., et al. (2001). The tumor suppressor PTEN positively regulates macroautophagy by inhibiting the phosphatidylinositol 3-kinase/protein kinase B pathway. *The Journal of Biological Chemistry*, *276*(38), 35243–35246. https://doi.org/10.1074/jbc.C100319200.

Bernhard, E. J., Stanbridge, E. J., Gupta, S., Gupta, A. K., Soto, D., Bakanauskas, V. J., et al. (2000). Direct evidence for the contribution of activated N-ras and K-ras oncogenes to increased intrinsic radiation resistance in human tumor cell lines. *Cancer Research*, *60*(23), 6597–6600.

Bhoopathi, P., Lee, N., Pradhan, A. K., Shen, X. N., Das, S. K., Sarkar, D., et al. (2016). mda-7/IL-24 induces cell death in neuroblastoma through a novel mechanism involving AIF and ATM. *Cancer Research*, *76*(12), 3572–3582. https://doi.org/10.1158/0008-5472.CAN-15-2959.

Bhutia, S. K., Dash, R., Das, S. K., Azab, B., Su, Z. Z., Lee, S. G., et al. (2010). Mechanism of autophagy to apoptosis switch triggered in prostate cancer cells by antitumor cytokine melanoma differentiation-associated gene 7/interleukin-24. *Cancer Research*, *70*(9), 3667–3676. https://doi.org/10.1158/0008-5472.CAN-09-3647.

Bhutia, S. K., Mukhopadhyay, S., Sinha, N., Das, D. N., Panda, P. K., Patra, S. K., et al. (2013). Autophagy: Cancer's friend or foe? *Advances in Cancer Research*, *118*, 61–95. https://doi.org/10.1016/B978-0-12-407173-5.00003-0.

Bonner, J. A., Raisch, K. P., Trummell, H. Q., Robert, F., Meredith, R. F., Spencer, S. A., et al. (2000). Enhanced apoptosis with combination C225/radiation treatment serves as the impetus for clinical investigation in head and neck cancers. *Journal of Clinical Oncology*, *18*(21 Suppl), 47S–53S.

Bristol, M. L., Di, X., Beckman, M. J., Wilson, E. N., Henderson, S. C., Maiti, A., et al. (2012). Dual functions of autophagy in the response of breast tumor cells to radiation:

Cytoprotective autophagy with radiation alone and cytotoxic autophagy in radiosensitization by vitamin D 3. *Autophagy*, *8*(5), 739–753. https://doi.org/10.4161/auto.19313.

Bristol, M. L., Emery, S. M., Maycotte, P., Thorburn, A., Chakradeo, S., & Gewirtz, D. A. (2013). Autophagy inhibition for chemosensitization and radiosensitization in cancer: Do the preclinical data support this therapeutic strategy? *The Journal of Pharmacology and Experimental Therapeutics*, *344*(3), 544–552. https://doi.org/10.1124/jpet.112.199802.

Cai-McRae, X., Zhong, H., & Karantza, V. (2015). Sequestosome 1/p62 facilitates HER2-induced mammary tumorigenesis through multiple signaling pathways. *Oncogene*, *34*(23), 2968–2977. https://doi.org/10.1038/onc.2014.244.

Cao, Y., & Klionsky, D. J. (2007). Physiological functions of Atg6/Beclin 1: A unique autophagy-related protein. *Cell Research*, *17*(10), 839–849. https://doi.org/10.1038/cr.2007.78.

Chada, S., Mhashilkar, A. M., Liu, Y., Nishikawa, T., Bocangel, D., Zheng, M., et al. (2006). mda-7 gene transfer sensitizes breast carcinoma cells to chemotherapy, biologic therapies and radiotherapy: Correlation with expression of bcl-2 family members. *Cancer Gene Therapy*, *13*(5), 490–502. https://doi.org/10.1038/sj.cgt.7700915.

Chakradeo, S., Sharma, K., Alhaddad, A., Bakhshwin, D., Le, N., Harada, H., et al. (2015). Yet another function of p53—The switch that determines whether radiation-induced autophagy will be cytoprotective or nonprotective: Implications for autophagy inhibition as a therapeutic strategy. *Molecular Pharmacology*, *87*(5), 803–814. https://doi.org/10.1124/mol.114.095273.

Cianfanelli, V., Fuoco, C., Lorente, M., Salazar, M., Quondamatteo, F., Gherardini, P. F., et al. (2015). AMBRA1 links autophagy to cell proliferation and tumorigenesis by promoting c-Myc dephosphorylation and degradation. *Nature Cell Biology*, *17*(1), 20–30. https://doi.org/10.1038/ncb3072.

Cliby, W., Ritland, S., Hartmann, L., Dodson, M., Halling, K. C., Keeney, G., et al. (1993). Human epithelial ovarian cancer allelotype. *Cancer Research*, *53*(10 Suppl), 2393–2398.

Copetti, T., Demarchi, F., & Schneider, C. (2009). p65/RelA binds and activates the beclin 1 promoter. *Autophagy*, *5*(6), 858–859.

Coppola, D., Helm, J., Ghayouri, M., Malafa, M. P., & Wang, H. G. (2011). Down-regulation of Bax-interacting factor 1 in human pancreatic ductal adenocarcinoma. *Pancreas*, *40*(3), 433–437. https://doi.org/10.1097/MPA.0b013e318205eb03.

Cuddeback, S. M., Yamaguchi, H., Komatsu, K., Miyashita, T., Yamada, M., Wu, C., et al. (2001). Molecular cloning and characterization of Bif-1. A novel Src homology 3 domain-containing protein that associates with Bax. *The Journal of Biological Chemistry*, *276*(23), 20559–20565. https://doi.org/10.1074/jbc.M101527200.

Cuervo, A. M. (2004). Autophagy: In sickness and in health. *Trends in Cell Biology*, *14*(2), 70–77. https://doi.org/10.1016/j.tcb.2003.12.002.

Cui, L., Song, Z., Liang, B., Jia, L., Ma, S., & Liu, X. (2016). Radiation induces autophagic cell death via the p53/DRAM signaling pathway in breast cancer cells. *Oncology Reports*, *35*(6), 3639–3647. https://doi.org/10.3892/or.2016.4752.

Cunningham, C. C., Chada, S., Merritt, J. A., Tong, A., Senzer, N., Zhang, Y., et al. (2005). Clinical and local biological effects of an intratumoral injection of mda-7 (IL24; INGN 241) in patients with advanced carcinoma: A phase I study. *Molecular Therapy*, *11*(1), 149–159. https://doi.org/10.1016/j.ymthe.2004.09.019.

Das, S. K., Sokhi, U. K., Bhutia, S. K., Azab, B., Su, Z. Z., Sarkar, D., et al. (2010). Human polynucleotide phosphorylase selectively and preferentially degrades microRNA-221 in human melanoma cells. *Proceedings of the National Academy of Sciences of the United States of America*, *107*(26), 11948–11953. https://doi.org/10.1073/pnas.0914143107.

Dash, R., Bhutia, S. K., Azab, B., Su, Z. Z., Quinn, B. A., Kegelmen, T. P., et al. (2010). mda-7/IL-24: A unique member of the IL-10 gene family promoting cancer-targeted toxicity. *Cytokine & Growth Factor Reviews, 21*(5), 381–391. https://doi.org/10.1016/j.cytogfr.2010.08.004.

De Amicis, F., Aquila, S., Morelli, C., Guido, C., Santoro, M., Perrotta, I., et al. (2015). Bergapten drives autophagy through the up-regulation of PTEN expression in breast cancer cells. *Molecular Cancer, 14*, 130. https://doi.org/10.1186/s12943-015-0403-4.

Ding, Z. B., Shi, Y. H., Zhou, J., Qiu, S. J., Xu, Y., Dai, Z., et al. (2008). Association of autophagy defect with a malignant phenotype and poor prognosis of hepatocellular carcinoma. *Cancer Research, 68*(22), 9167–9175. https://doi.org/10.1158/0008-5472.CAN-08-1573.

Eccles, D. M., Russell, S. E., Haites, N. E., Atkinson, R., Bell, D. W., Gruber, L., et al. (1992). Early loss of heterozygosity on 17q in ovarian cancer. The Abe ovarian cancer genetics group. *Oncogene, 7*(10), 2069–2072.

Emdad, L., Lebedeva, I. V., Su, Z. Z., Gupta, P., Sarkar, D., Settleman, J., et al. (2007). Combinatorial treatment of non-small-cell lung cancers with gefitinib and Ad.mda-7 enhances apoptosis-induction and reverses resistance to a single therapy. *Journal of Cellular Physiology, 210*(2), 549–559. https://doi.org/10.1002/jcp.20906.

Emdad, L., Lebedeva, I. V., Su, Z. Z., Gupta, P., Sauane, M., Dash, R., et al. (2009). Historical perspective and recent insights into our understanding of the molecular and biochemical basis of the antitumor properties of mda-7/IL-24. *Cancer Biology & Therapy, 8*(5), 391–400.

Emdad, L., Lebedeva, I. V., Su, Z. Z., Sarkar, D., Dent, P., Curiel, D. T., et al. (2007). Melanoma differentiation associated gene-7/interleukin-24 reverses multidrug resistance in human colorectal cancer cells. *Molecular Cancer Therapeutics, 6*(11), 2985–2994. https://doi.org/10.1158/1535-7163.MCT-07-0399.

Fimia, G. M., Corazzari, M., Antonioli, M., & Piacentini, M. (2013). Ambra1 at the crossroad between autophagy and cell death. *Oncogene, 32*(28), 3311–3318. https://doi.org/10.1038/onc.2012.455.

Fimia, G. M., Stoykova, A., Romagnoli, A., Giunta, L., Di Bartolomeo, S., Nardacci, R., et al. (2007). Ambra1 regulates autophagy and development of the nervous system. *Nature, 447*(7148), 1121–1125. https://doi.org/10.1038/nature05925.

Finka, A., & Goloubinoff, P. (2013). Proteomic data from human cell cultures refine mechanisms of chaperone-mediated protein homeostasis. *Cell Stress & Chaperones, 18*(5), 591–605. https://doi.org/10.1007/s12192-013-0413-3.

Fisher, P. B. (2005). Is mda-7/IL-24 a "magic bullet" for cancer? *Cancer Research, 65*(22), 10128–10138. https://doi.org/10.1158/0008-5472.CAN-05-3127.

Fisher, P. B., Gopalkrishnan, R. V., Chada, S., Ramesh, R., Grimm, E. A., Rosenfeld, M. R., et al. (2003). mda-7/IL-24, a novel cancer selective apoptosis inducing cytokine gene: From the laboratory into the clinic. *Cancer Biology & Therapy, 2*(4 Suppl. 1), S23–37.

Fisher, P. B., Sarkar, D., Lebedeva, I. V., Emdad, L., Gupta, P., Sauane, M., et al. (2007). Melanoma differentiation associated gene-7/interleukin-24 (mda-7/IL-24): Novel gene therapeutic for metastatic melanoma. *Toxicology and Applied Pharmacology, 224*(3), 300–307. https://doi.org/10.1016/j.taap.2006.11.021.

Fornari, F., Gramantieri, L., Ferracin, M., Veronese, A., Sabbioni, S., Calin, G. A., et al. (2008). MiR-221 controls CDKN1C/p57 and CDKN1B/p27 expression in human hepatocellular carcinoma. *Oncogene, 27*(43), 5651–5661. https://doi.org/10.1038/onc.2008.178.

Fujiwara, N., Usui, T., Ohama, T., & Sato, K. (2016). Regulation of Beclin 1 protein phosphorylation and autophagy by protein phosphatase 2A (PP2A) and death-associated protein kinase 3 (DAPK3). *The Journal of Biological Chemistry*, *291*(20), 10858–10866. https://doi.org/10.1074/jbc.M115.704908.

Furuya, N., Yu, J., Byfield, M., Pattingre, S., & Levine, B. (2005). The evolutionarily conserved domain of Beclin 1 is required for Vps34 binding, autophagy and tumor suppressor function. *Autophagy*, *1*(1), 46–52.

Futreal, P. A., Soderkvist, P., Marks, J. R., Iglehart, J. D., Cochran, C., Barrett, J. C., et al. (1992). Detection of frequent allelic loss on proximal chromosome 17q in sporadic breast carcinoma using microsatellite length polymorphisms. *Cancer Research*, *52*(9), 2624–2627.

Gallop, J. L., Jao, C. C., Kent, H. M., Butler, P. J., Evans, P. R., Langen, R., et al. (2006). Mechanism of endophilin N-BAR domain-mediated membrane curvature. *The EMBO Journal*, *25*(12), 2898–2910. https://doi.org/10.1038/sj.emboj.7601174.

Gao, P., Sun, X., Chen, X., Wang, Y., Foster, B. A., Subjeck, J., et al. (2008). Secretable chaperone Grp170 enhances therapeutic activity of a novel tumor suppressor, mda-7/IL-24. *Cancer Research*, *68*(10), 3890–3898. https://doi.org/10.1158/0008-5472.CAN-08-0156.

Gao, X., Zacharek, A., Salkowski, A., Grignon, D. J., Sakr, W., Porter, A. T., et al. (1995). Loss of heterozygosity of the BRCA1 and other loci on chromosome 17q in human prostate cancer. *Cancer Research*, *55*(5), 1002–1005.

Garofalo, M., Di Leva, G., Romano, G., Nuovo, G., Suh, S. S., Ngankeu, A., et al. (2009). miR-221&222 regulate TRAIL resistance and enhance tumorigenicity through PTEN and TIMP3 downregulation. *Cancer Cell*, *16*(6), 498–509. https://doi.org/10.1016/j.ccr.2009.10.014.

Gawriluk, T. R., Ko, C., Hong, X., Christenson, L. K., & Rucker, E. B., 3rd (2014). Beclin-1 deficiency in the murine ovary results in the reduction of progesterone production to promote preterm labor. *Proceedings of the National Academy of Sciences of the United States of America*, *111*(40), E4194–4203. https://doi.org/10.1073/pnas.1409323111.

Gewirtz, D. A. (2014). The autophagic response to radiation: Relevance for radiation sensitization in cancer therapy. *Radiation Research*, *182*(4), 363–367. https://doi.org/10.1667/RR13774.1.

Giles, N. M., Gutowski, N. J., Giles, G. I., & Jacob, C. (2003). Redox catalysts as sensitisers towards oxidative stress. *FEBS Letters*, *535*(1–3), 179–182.

Gong, C., Bauvy, C., Tonelli, G., Yue, W., Delomenie, C., Nicolas, V., et al. (2013). Beclin 1 and autophagy are required for the tumorigenicity of breast cancer stem-like/progenitor cells. *Oncogene*, *32*(18), 2261–2272. 2272e.2261–2211. https://doi.org/10.1038/onc.2012.252.

Gu, W., Wan, D., Qian, Q., Yi, B., He, Z., Gu, Y., et al. (2014). Ambra1 is an essential regulator of autophagy and apoptosis in SW620 cells: Pro-survival role of Ambra1. *PLoS One*, *9*(2), e90151. https://doi.org/10.1371/journal.pone.0090151.

Guo, G. F., Jiang, W. Q., Zhang, B., Cai, Y. C., Xu, R. H., Chen, X. X., et al. (2011). Autophagy-related proteins Beclin-1 and LC3 predict cetuximab efficacy in advanced colorectal cancer. *World Journal of Gastroenterology*, *17*(43), 4779–4786. https://doi.org/10.3748/wjg.v17.i43.4779.

Gupta, P., Walter, M. R., Su, Z. Z., Lebedeva, I. V., Emdad, L., Randolph, A., et al. (2006). BiP/GRP78 is an intracellular target for MDA-7/IL-24 induction of cancer-specific apoptosis. *Cancer Research*, *66*(16), 8182–8191. https://doi.org/10.1158/0008-5472.CAN-06-0577.

Han, J., Hou, W., Lu, C., Goldstein, L. A., Stolz, D. B., Watkins, S. C., et al. (2013). Interaction between Her2 and Beclin-1 proteins underlies a new mechanism of reciprocal regulation. *The Journal of Biological Chemistry*, *288*(28), 20315–20325. https://doi.org/10.1074/jbc.M113.461350.

Han, Y., Xue, X. F., Shen, H. G., Guo, X. B., Wang, X., Yuan, B., et al. (2014). Prognostic significance of Beclin-1 expression in colorectal cancer: A meta-analysis. *Asian Pacific Journal of Cancer Prevention*, *15*(11), 4583–4587.

Hatziapostolou, M., Polytarchou, C., Aggelidou, E., Drakaki, A., Poultsides, G. A., Jaeger, S. A., et al. (2011). An HNF4alpha-miRNA inflammatory feedback circuit regulates hepatocellular oncogenesis. *Cell*, *147*(6), 1233–1247. https://doi.org/10.1016/j.cell.2011.10.043.

He, W. S., Dai, X. F., Jin, M., Liu, C. W., & Rent, J. H. (2012). Hypoxia-induced autophagy confers resistance of breast cancer cells to ionizing radiation. *Oncology Research*, *20*(5–6), 251–258.

Ho, J., Kong, J. W., Choong, L. Y., Loh, M. C., Toy, W., Chong, P. K., et al. (2009). Novel breast cancer metastasis-associated proteins. *Journal of Proteome Research*, *8*(2), 583–594. https://doi.org/10.1021/pr8007368.

Huang, J. J., Li, H. R., Huang, Y., Jiang, W. Q., Xu, R. H., Huang, H. Q., et al. (2010). Beclin 1 expression: A predictor of prognosis in patients with extranodal natural killer T-cell lymphoma, nasal type. *Autophagy*, *6*(6), 777–783.

Huang, X., Qi, Q., Hua, X., Li, X., Zhang, W., Sun, H., et al. (2014). Beclin 1, an autophagy-related gene, augments apoptosis in U87 glioblastoma cells. *Oncology Reports*, *31*(4), 1761–1767. https://doi.org/10.3892/or.2014.3015.

Huettenbrenner, S., Maier, S., Leisser, C., Polgar, D., Strasser, S., Grusch, M., et al. (2003). The evolution of cell death programs as prerequisites of multicellularity. *Mutation Research*, *543*(3), 235–249.

Jiang, H., & Fisher, P. B. (1993). Use of a sensitive and efficient subtraction hybridization protocol for the identification of genes differentially regulated during the induction of differentiation in human melanoma cells. *Molecular and Cellular Differentiation*, *1*(3), 285–299.

Jiang, H., Lin, J. J., Su, Z. Z., Goldstein, N. I., & Fisher, P. B. (1995). Subtraction hybridization identifies a novel melanoma differentiation associated gene, mda-7, modulated during human melanoma differentiation, growth and progression. *Oncogene*, *11*(12), 2477–2486.

Kim, Y. M., Jung, C. H., Seo, M., Kim, E. K., Park, J. M., Bae, S. S., et al. (2015). mTORC1 phosphorylates UVRAG to negatively regulate autophagosome and endosome maturation. *Molecular Cell*, *57*(2), 207–218. https://doi.org/10.1016/j.molcel.2014.11.013.

Kim, J., Kundu, M., Viollet, B., & Guan, K. L. (2011). AMPK and mTOR regulate autophagy through direct phosphorylation of Ulk1. *Nature Cell Biology*, *13*(2), 132–141. https://doi.org/10.1038/ncb2152.

Kim, M. S., Yoo, N. J., & Lee, S. H. (2008). Somatic mutation of pro-cell death Bif-1 gene is rare in common human cancers. *APMIS*, *116*(10), 939–940. https://doi.org/10.1111/j.1600-0463.2008.01091.x.

Klein, S. R., Piya, S., Lu, Z., Xia, Y., Alonso, M. M., White, E. J., et al. (2015). C-Jun N-terminal kinases are required for oncolytic adenovirus-mediated autophagy. *Oncogene*, *34*, 5295–5301. https://doi.org/10.1038/onc.2014.452.

Klionsky, D. J. (2007). Autophagy: From phenomenology to molecular understanding in less than a decade. *Nature Reviews. Molecular Cell Biology*, *8*(11), 931–937. https://doi.org/10.1038/nrm2245.

Knudson, A. G., Jr. (1971). Mutation and cancer: Statistical study of retinoblastoma. *Proceedings of the National Academy of Sciences of the United States of America*, *68*(4), 820–823.

Ko, Y. H., Cho, Y. S., Won, H. S., Jeon, E. K., An, H. J., Hong, S. U., et al. (2013). Prognostic significance of autophagy-related protein expression in resected pancreatic ductal adenocarcinoma. *Pancreas*, *42*(5), 829–835. https://doi.org/10.1097/MPA.0b013e318279d0dc.

Koukourakis, M. I., Giatromanolaki, A., Sivridis, E., Pitiakoudis, M., Gatter, K. C., & Harris, A. L. (2010). Beclin 1 over- and underexpression in colorectal cancer: Distinct patterns relate to prognosis and tumour hypoxia. *British Journal of Cancer, 103*(8), 1209–1214. https://doi.org/10.1038/sj.bjc.6605904.

Kroemer, G., Galluzzi, L., Vandenabeele, P., Abrams, J., Alnemri, E. S., Baehrecke, E. H., et al. (2009). Classification of cell death: Recommendations of the nomenclature committee on cell death 2009. *Cell Death and Differentiation, 16*(1), 3–11. https://doi.org/10.1038/cdd.2008.150.

Kung, C. P., Budina, A., Balaburski, G., Bergenstock, M. K., & Murphy, M. (2011). Autophagy in tumor suppression and cancer therapy. *Critical Reviews in Eukaryotic Gene Expression, 21*(1), 71–100.

le Sage, C., Nagel, R., Egan, D. A., Schrier, M., Mesman, E., Mangiola, A., et al. (2007). Regulation of the p27(Kip1) tumor suppressor by miR-221 and miR-222 promotes cancer cell proliferation. *The EMBO Journal, 26*(15), 3699–3708. https://doi.org/10.1038/sj.emboj.7601790.

Lebedeva, I. V., Sauane, M., Gopalkrishnan, R. V., Sarkar, D., Su, Z. Z., Gupta, P., et al. (2005). mda-7/IL-24: Exploiting cancer's Achilles' heel. *Molecular Therapy, 11*(1), 4–18. https://doi.org/10.1016/j.ymthe.2004.08.012.

Lee, J., Giordano, S., & Zhang, J. (2012). Autophagy, mitochondria and oxidative stress: Cross-talk and redox signalling. *The Biochemical Journal, 441*(2), 523–540. https://doi.org/10.1042/BJ20111451.

Lee, S. J., Kim, H. P., Jin, Y., Choi, A. M., & Ryter, S. W. (2011). Beclin 1 deficiency is associated with increased hypoxia-induced angiogenesis. *Autophagy, 7*(8), 829–839.

Levine, B., & Klionsky, D. J. (2004). Development by self-digestion: Molecular mechanisms and biological functions of autophagy. *Developmental Cell, 6*(4), 463–477.

Li, Z., Chen, B., Wu, Y., Jin, F., Xia, Y., & Liu, X. (2010). Genetic and epigenetic silencing of the beclin 1 gene in sporadic breast tumors. *BMC Cancer, 10*, 98. https://doi.org/10.1186/1471-2407-10-98.

Li, B. X., Li, C. Y., Peng, R. Q., Wu, X. J., Wang, H. Y., Wan, D. S., et al. (2009). The expression of beclin 1 is associated with favorable prognosis in stage IIIB colon cancers. *Autophagy, 5*(3), 303–306.

Li, X., Lu, Y., Pan, T., & Fan, Z. (2010). Roles of autophagy in cetuximab-mediated cancer therapy against EGFR. *Autophagy, 6*(8), 1066–1077. https://doi.org/10.4161/auto.6.8.13366.

Li, X., Yan, J., Wang, L., Xiao, F., Yang, Y., Guo, X., et al. (2013). Beclin1 inhibition promotes autophagy and decreases gemcitabine-induced apoptosis in Miapaca2 pancreatic cancer cells. *Cancer Cell International, 13*(1), 26. https://doi.org/10.1186/1475-2867-13-26.

Li, Y., Zhang, J., Chen, X., Liu, T., He, W., Chen, Y., et al. (2012). Molecular machinery of autophagy and its implication in cancer. *The American Journal of the Medical Sciences, 343*(2), 155–161. https://doi.org/10.1097/MAJ.0b013e31821f978d.

Liang, X. H., Jackson, S., Seaman, M., Brown, K., Kempkes, B., Hibshoosh, H., et al. (1999). Induction of autophagy and inhibition of tumorigenesis by beclin 1. *Nature, 402*(6762), 672–676. https://doi.org/10.1038/45257.

Liang, X. H., Kleeman, L. K., Jiang, H. H., Gordon, G., Goldman, J. E., Berry, G., et al. (1998). Protection against fatal Sindbis virus encephalitis by beclin, a novel Bcl-2-interacting protein. *Journal of Virology, 72*(11), 8586–8596.

Liang, X. H., Yu, J., Brown, K., & Levine, B. (2001). Beclin 1 contains a leucine-rich nuclear export signal that is required for its autophagy and tumor suppressor function. *Cancer Research, 61*(8), 3443–3449.

Lin, H. X., Qiu, H. J., Zeng, F., Rao, H. L., Yang, G. F., Kung, H. F., et al. (2013). Decreased expression of Beclin 1 correlates closely with Bcl-xL expression and poor prognosis of ovarian carcinoma. *PLoS One, 8*(4), e60516. https://doi.org/10.1371/journal.pone.0060516.

Liu, J., & Debnath, J. (2016). The evolving, multifaceted roles of autophagy in cancer. *Advances in Cancer Research, 130,* 1–53. https://doi.org/10.1016/bs.acr.2016.01.005.

Ma, K., Zhang, C., Huang, M. Y., Guo, Y. X., & Hu, G. Q. (2016). Crosstalk between Beclin-1-dependent autophagy and caspasedependent apoptosis induced by tanshinone IIA in human osteosarcoma MG-63 cells. *Oncology Reports, 36*(4), 1807–1818. https://doi.org/10.3892/or.2016.5003.

Maddison, W. P., & Maddison, D. R. (2017). *Mesquite: A modular system for evolutionary analysis.* Version 3.2. http://mesquiteproject.org.

Martin, M., Hinojar, A., Cerezo, L., Garcia, J., Lopez, M., Prada, J., et al. (2016). Aldehyde dehydrogenase isoform 1 (ALDH1) expression as a predictor of radiosensitivity in laryngeal cancer. *Clinical & Translational Oncology, 18*(8), 825–830. https://doi.org/10.1007/s12094-015-1445-1.

Masuda, M., Takeda, S., Sone, M., Ohki, T., Mori, H., Kamioka, Y., et al. (2006). Endophilin BAR domain drives membrane curvature by two newly identified structure-based mechanisms. *The EMBO Journal, 25*(12), 2889–2897. https://doi.org/10.1038/sj.emboj.7601176.

Mathew, R., Karp, C. M., Beaudoin, B., Vuong, N., Chen, G., Chen, H. Y., et al. (2009). Autophagy suppresses tumorigenesis through elimination of p62. *Cell, 137*(6), 1062–1075. https://doi.org/10.1016/j.cell.2009.03.048.

Mathew, R., & White, E. (2011). Autophagy in tumorigenesis and energy metabolism: Friend by day, foe by night. *Current Opinion in Genetics & Development, 21*(1), 113–119. https://doi.org/10.1016/j.gde.2010.12.008.

Miracco, C., Cevenini, G., Franchi, A., Luzi, P., Cosci, E., Mourmouras, V., et al. (2010). Beclin 1 and LC3 autophagic gene expression in cutaneous melanocytic lesions. *Human Pathology, 41*(4), 503–512. https://doi.org/10.1016/j.humpath.2009.09.004.

Miracco, C., Cosci, E., Oliveri, G., Luzi, P., Pacenti, L., Monciatti, I., et al. (2007). Protein and mRNA expression of autophagy gene Beclin 1 in human brain tumours. *International Journal of Oncology, 30*(2), 429–436.

Mizushima, N. (2007). Autophagy: Process and function. *Genes & Development, 21*(22), 2861–2873. https://doi.org/10.1101/gad.1599207.

Mizushima, N., Levine, B., Cuervo, A. M., & Klionsky, D. J. (2008). Autophagy fights disease through cellular self-digestion. *Nature, 451*(7182), 1069–1075. https://doi.org/10.1038/nature06639.

Mohan, C. D., Srinivasa, V., Rangappa, S., Mervin, L., Mohan, S., Paricharak, S., et al. (2016). Trisubstituted-imidazoles induce apoptosis in human breast cancer cells by targeting the oncogenic PI3K/Akt/mTOR signaling pathway. *PLoS One, 11*(4), e0153155. https://doi.org/10.1371/journal.pone.0153155.

Moreau, K., Luo, S., & Rubinsztein, D. C. (2010). Cytoprotective roles for autophagy. *Current Opinion in Cell Biology, 22*(2), 206–211. https://doi.org/10.1016/j.ceb.2009.12.002.

Morran, D. C., Wu, J., Jamieson, N. B., Mrowinska, A., Kalna, G., Karim, S. A., et al. (2014). Targeting mTOR dependency in pancreatic cancer. *Gut, 63*(9), 1481–1489. https://doi.org/10.1136/gutjnl-2013-306202.

Mowers, E. E., Sharifi, M. N., & Macleod, K. F. (2017). Autophagy in cancer metastasis. *Oncogene, 36*(12), 1619–1630. https://doi.org/10.1038/onc.2016.333.

Nazio, F., Strappazzon, F., Antonioli, M., Bielli, P., Cianfanelli, V., Bordi, M., et al. (2013). mTOR inhibits autophagy by controlling ULK1 ubiquitylation, self-association and function through AMBRA1 and TRAF6. *Nature Cell Biology, 15*(4), 406–416. https://doi.org/10.1038/ncb2708.

Nishikawa, T., Ramesh, R., Munshi, A., Chada, S., & Meyn, R. E. (2004). Adenovirus-mediated mda-7 (IL24) gene therapy suppresses angiogenesis and sensitizes NSCLC xenograft tumors to radiation. *Molecular Therapy*, *9*(6), 818–828. https://doi.org/10.1016/j.ymthe.2004.03.014.

Noble, C. G., Dong, J. M., Manser, E., & Song, H. (2008). Bcl-xL and UVRAG cause a monomer-dimer switch in Beclin1. *The Journal of Biological Chemistry*, *283*(38), 26274–26282. https://doi.org/10.1074/jbc.M804723200.

Park, J. M., Tougeron, D., Huang, S., Okamoto, K., & Sinicrope, F. A. (2014). Beclin 1 and UVRAG confer protection from radiation-induced DNA damage and maintain centrosome stability in colorectal cancer cells. *PLoS One*, *9*(6), e100819. https://doi.org/10.1371/journal.pone.0100819.

Peter, B. J., Kent, H. M., Mills, I. G., Vallis, Y., Butler, P. J., Evans, P. R., et al. (2004). BAR domains as sensors of membrane curvature: The amphiphysin BAR structure. *Science*, *303*(5657), 495–499. https://doi.org/10.1126/science.1092586.

Petiot, A., Ogier-Denis, E., Blommaart, E. F., Meijer, A. J., & Codogno, P. (2000). Distinct classes of phosphatidylinositol 3'-kinases are involved in signaling pathways that control macroautophagy in HT-29 cells. *The Journal of Biological Chemistry*, *275*(2), 992–998.

Pierrat, B., Simonen, M., Cueto, M., Mestan, J., Ferrigno, P., & Heim, J. (2001). SH3GLB, a new endophilin-related protein family featuring an SH3 domain. *Genomics*, *71*(2), 222–234. https://doi.org/10.1006/geno.2000.6378.

Pirtoli, L., Cevenini, G., Tini, P., Vannini, M., Oliveri, G., Marsili, S., et al. (2009). The prognostic role of Beclin 1 protein expression in high-grade gliomas. *Autophagy*, *5*(7), 930–936.

Pradhan, A. K., Talukdar, S., Bhoopathi, P., Shen, X. N., Emdad, L., Das, S. K., et al. (2017). mda-7/IL-24 mediates cancer cell-specific death via regulation of miR-221 and the Beclin-1 Axis. *Cancer Research*, *77*(4), 949–959. https://doi.org/10.1158/0008-5472.CAN-16-1731.

Qu, X., Yu, J., Bhagat, G., Furuya, N., Hibshoosh, H., Troxel, A., et al. (2003). Promotion of tumorigenesis by heterozygous disruption of the beclin 1 autophagy gene. *The Journal of Clinical Investigation*, *112*(12), 1809–1820. https://doi.org/10.1172/JCI20039.

Radwan, S. M., Hamdy, N. M., Hegab, H. M., & El-Mesallamy, H. O. (2016). Beclin-1 and hypoxia-inducible factor-1alpha genes expression: Potential biomarkers in acute leukemia patients. *Cancer Biomarkers*, *16*(4), 619–626. https://doi.org/10.3233/CBM-160603.

Ramesh, R., Mhashilkar, A. M., Tanaka, F., Saito, Y., Branch, C. D., Sieger, K., et al. (2003). Melanoma differentiation-associated gene 7/interleukin (IL)-24 is a novel ligand that regulates angiogenesis via the IL-22 receptor. *Cancer Research*, *63*(16), 5105–5113.

Ren, J., & Taegtmeyer, H. (2015). Too much or not enough of a good thing—The Janus faces of autophagy in cardiac fuel and protein homeostasis. *Journal of Molecular and Cellular Cardiology*, *84*, 223–226. https://doi.org/10.1016/j.yjmcc.2015.03.001.

Riley, B. E., Kaiser, S. E., Shaler, T. A., Ng, A. C., Hara, T., Hipp, M. S., et al. (2010). Ubiquitin accumulation in autophagy-deficient mice is dependent on the Nrf2-mediated stress response pathway: A potential role for protein aggregation in autophagic substrate selection. *The Journal of Cell Biology*, *191*(3), 537–552. https://doi.org/10.1083/jcb.201005012.

Rosenfeld, J. A., & DeSalle, R. (2012). E value cutoff and eukaryotic genome content phylogenetics. *Molecular Phylogenetics and Evolution*, *63*(2), 342–350.

Runkle, K. B., Meyerkord, C. L., Desai, N. V., Takahashi, Y., & Wang, H. G. (2012). Bif-1 suppresses breast cancer cell migration by promoting EGFR endocytic degradation. *Cancer Biology & Therapy*, *13*(10), 956–966. https://doi.org/10.4161/cbt.20951.

Russell, S. E., Hickey, G. I., Lowry, W. S., White, P., & Atkinson, R. J. (1990). Allele loss from chromosome 17 in ovarian cancer. *Oncogene*, *5*(10), 1581–1583.

Saito, H., Inazawa, J., Saito, S., Kasumi, F., Koi, S., Sagae, S., et al. (1993). Detailed deletion mapping of chromosome 17q in ovarian and breast cancers: 2-cM region on 17q21.3 often and commonly deleted in tumors. *Cancer Research, 53*(14), 3382–3385.

Sarkar, D., Su, Z. Z., Lebedeva, I. V., Sauane, M., Gopalkrishnan, R. V., Valerie, K., et al. (2002). mda-7 (IL-24) mediates selective apoptosis in human melanoma cells by inducing the coordinated overexpression of the GADD family of genes by means of p38 MAPK. *Proceedings of the National Academy of Sciences of the United States of America, 99*(15), 10054–10059. https://doi.org/10.1073/pnas.152327199.

Sauane, M., Su, Z. Z., Gupta, P., Lebedeva, I. V., Dent, P., Sarkar, D., et al. (2008). Autocrine regulation of mda-7/IL-24 mediates cancer-specific apoptosis. *Proceedings of the National Academy of Sciences of the United States of America, 105*(28), 9763–9768. https://doi.org/10.1073/pnas.0804089105.

Shin, J. Y., Lim, H. T., Minai-Tehrani, A., Noh, M. S., Kim, J. E., Kim, J. H., et al. (2012). Aerosol delivery of beclin1 enhanced the anti-tumor effect of radiation in the lungs of K-rasLA1 mice. *Journal of Radiation Research, 53*(4), 506–515. https://doi.org/10.1093/jrr/rrs005.

Singh, S. P., Pandey, T., Srivastava, R., Verma, P. C., Singh, P. K., Tuli, R., et al. (2010). BECLIN1 from Arabidopsis thaliana under the generic control of regulated expression systems, a strategy for developing male sterile plants. *Plant Biotechnology Journal, 8*(9), 1005–1022. https://doi.org/10.1111/j.1467-7652.2010.00527.x.

Su, Z., Emdad, L., Sauane, M., Lebedeva, I. V., Sarkar, D., Gupta, P., et al. (2005). Unique aspects of mda-7/IL-24 antitumor bystander activity: Establishing a role for secretion of MDA-7/IL-24 protein by normal cells. *Oncogene, 24*(51), 7552–7566. https://doi.org/10.1038/sj.onc.1208911.

Su, Z. Z., Lebedeva, I. V., Sarkar, D., Gopalkrishnan, R. V., Sauane, M., Sigmon, C., et al. (2003). Melanoma differentiation associated gene-7, mda-7/IL-24, selectively induces growth suppression, apoptosis and radiosensitization in malignant gliomas in a p53-independent manner. *Oncogene, 22*(8), 1164–1180. https://doi.org/10.1038/sj.onc.1206062.

Su, Z. Z., Madireddi, M. T., Lin, J. J., Young, C. S., Kitada, S., Reed, J. C., et al. (1998). The cancer growth suppressor gene mda-7 selectively induces apoptosis in human breast cancer cells and inhibits tumor growth in nude mice. *Proceedings of the National Academy of Sciences of the United States of America, 95*(24), 14400–14405.

Sun, W. L. (2016). Ambra1 in autophagy and apoptosis: Implications for cell survival and chemotherapy resistance. *Oncology Letters, 12*(1), 367–374. https://doi.org/10.3892/ol.2016.4644.

Sun, Y., Liu, J. H., Jin, L., Lin, S. M., Yang, Y., Sui, Y. X., et al. (2010). Over-expression of the Beclin1 gene upregulates chemosensitivity to anti-cancer drugs by enhancing therapy-induced apoptosis in cervix squamous carcinoma CaSki cells. *Cancer Letters, 294*(2), 204–210. https://doi.org/10.1016/j.canlet.2010.02.001.

Sun, Y., Liu, J. H., Sui, Y. X., Jin, L., Yang, Y., Lin, S. M., et al. (2011). Beclin1 overexpression inhibits proliferation, invasion and migration of CaSki cervical cancer cells. *Asian Pacific Journal of Cancer Prevention, 12*(5), 1269–1273.

Takahashi, Y., Coppola, D., Matsushita, N., Cualing, H. D., Sun, M., Sato, Y., et al. (2007). Bif-1 interacts with Beclin 1 through UVRAG and regulates autophagy and tumorigenesis. *Nature Cell Biology, 9*(10), 1142–1151. https://doi.org/10.1038/ncb1634.

Tan, X., Thapa, N., Sun, Y., & Anderson, R. A. (2015). A kinase-independent role for EGF receptor in autophagy initiation. *Cell, 160*(1–2), 145–160. https://doi.org/10.1016/j.cell.2014.12.006.

Tan, Q., Wang, M., Yu, M., Zhang, J., Bristow, R. G., Hill, R. P., et al. (2016). Role of autophagy as a survival mechanism for hypoxic cells in tumors. *Neoplasia, 18*(6), 347–355. https://doi.org/10.1016/j.neo.2016.04.003.

Tangir, J., Muto, M. G., Berkowitz, R. S., Welch, W. R., Bell, D. A., & Mok, S. C. (1996). A 400 kb novel deletion unit centromeric to the BRCA1 gene in sporadic epithelial ovarian cancer. *Oncogene, 12*(4), 735–740.

Tong, A. W., Nemunaitis, J., Su, D., Zhang, Y., Cunningham, C., Senzer, N., et al. (2005). Intratumoral injection of INGN 241, a nonreplicating adenovector expressing the melanoma-differentiation associated gene-7 (mda-7/IL24): Biologic outcome in advanced cancer patients. *Molecular Therapy, 11*(1), 160–172. https://doi.org/10.1016/j.ymthe.2004.09.021.

Valente, G., Morani, F., Nicotra, G., Fusco, N., Peracchio, C., Titone, R., et al. (2014). Expression and clinical significance of the autophagy proteins BECLIN 1 and LC3 in ovarian cancer. *BioMed Research International, 2014*, 462658. https://doi.org/10.1155/2014/462658.

Wang, B., Ling, S., & Lin, W. C. (2010). 14-3-3Tau regulates Beclin 1 and is required for autophagy. *PLoS One, 5*(4), e10409. https://doi.org/10.1371/journal.pone.0010409.

Wang, R. C., Wei, Y., An, Z., Zou, Z., Xiao, G., Bhagat, G., et al. (2012). Akt-mediated regulation of autophagy and tumorigenesis through Beclin 1 phosphorylation. *Science, 338*(6109), 956–959. https://doi.org/10.1126/science.1225967.

Wei, Y., Pattingre, S., Sinha, S., Bassik, M., & Levine, B. (2008). JNK1-mediated phosphorylation of Bcl-2 regulates starvation-induced autophagy. *Molecular Cell, 30*(6), 678–688. https://doi.org/10.1016/j.molcel.2008.06.001.

Wei, Y., Zou, Z., Becker, N., Anderson, M., Sumpter, R., Xiao, G., et al. (2013). EGFR-mediated Beclin 1 phosphorylation in autophagy suppression, tumor progression, and tumor chemoresistance. *Cell, 154*(6), 1269–1284. https://doi.org/10.1016/j.cell.2013.08.015.

White, E., & DiPaola, R. S. (2009). The double-edged sword of autophagy modulation in cancer. *Clinical Cancer Research, 15*(17), 5308–5316. https://doi.org/10.1158/1078-0432.CCR-07-5023.

Wilson, E. N., Bristol, M. L., Di, X., Maltese, W. A., Koterba, K., Beckman, M. J., et al. (2011). A switch between cytoprotective and cytotoxic autophagy in the radiosensitization of breast tumor cells by chloroquine and vitamin D. *Hormones and Cancer, 2*(5), 272–285. https://doi.org/10.1007/s12672-011-0081-7.

Wirawan, E., Lippens, S., Vanden Berghe, T., Romagnoli, A., Fimia, G. M., Piacentini, M., et al. (2012). Beclin1: A role in membrane dynamics and beyond. *Autophagy, 8*(1), 6–17. https://doi.org/10.4161/auto.8.1.16645.

Xu, F., Fang, Y., Yan, L., Xu, L., Zhang, S., Cao, Y., et al. (2017). Nuclear localization of Beclin 1 promotes radiation-induced DNA damage repair independent of autophagy. *Scientific Reports, 7*, 45385. https://doi.org/10.1038/srep45385.

Yacoub, A., Mitchell, C., Lister, A., Lebedeva, I. V., Sarkar, D., Su, Z. Z., et al. (2003). Melanoma differentiation-associated 7 (interleukin 24) inhibits growth and enhances radiosensitivity of glioma cells in vitro and in vivo. *Clinical Cancer Research, 9*(9), 3272–3281.

Yamaguchi, H., Woods, N. T., Dorsey, J. F., Takahashi, Y., Gjertsen, N. R., Yeatman, T., et al. (2008). SRC directly phosphorylates Bif-1 and prevents its interaction with Bax and the initiation of anoikis. *The Journal of Biological Chemistry, 283*(27), 19112–19118. https://doi.org/10.1074/jbc.M709882200.

Yang, W., Ju, J. H., Lee, K. M., Nam, K., Oh, S., & Shin, I. (2013). Protein kinase B/Akt1 inhibits autophagy by down-regulating UVRAG expression. *Experimental Cell Research, 319*(3), 122–133. https://doi.org/10.1016/j.yexcr.2012.11.014.

Yin, X., Cao, L., Peng, Y., Tan, Y., Xie, M., Kang, R., et al. (2011). A critical role for UVRAG in apoptosis. *Autophagy, 7*(10), 1242–1244. https://doi.org/10.4161/auto.7.10.16507.

Ying, H., Qu, D., Liu, C., Ying, T., Lv, J., Jin, S., et al. (2015). Chemoresistance is associated with Beclin-1 and PTEN expression in epithelial ovarian cancers. *Oncology Letters*, *9*(4), 1759–1763. https://doi.org/10.3892/ol.2015.2950.

Zhang, Z., Shao, Z., Xiong, L., Che, B., Deng, C., & Xu, W. (2009). Expression of Beclin1 in osteosarcoma and the effects of down-regulation of autophagy on the chemotherapeutic sensitivity. *Journal of Huazhong University of Science and Technology Medical Sciences*, *29*(6), 737–740. https://doi.org/10.1007/s11596-009-0613-3.

Zhang, D., Wang, W., Sun, X., Xu, D., Wang, C., Zhang, Q., et al. (2016). AMPK regulates autophagy by phosphorylating BECN1 at threonine 388. *Autophagy*, *12*(9), 1447–1459. https://doi.org/10.1080/15548627.2016.1185576.

Zhang, C., Zhang, J., Zhang, A., Wang, Y., Han, L., You, Y., et al. (2010). PUMA is a novel target of miR-221/222 in human epithelial cancers. *International Journal of Oncology*, *37*(6), 1621–1626.

Zhao, Z., Oh, S., Li, D., Ni, D., Pirooz, S. D., Lee, J. H., et al. (2012). A dual role for UVRAG in maintaining chromosomal stability independent of autophagy. *Developmental Cell*, *22*(5), 1001–1016. https://doi.org/10.1016/j.devcel.2011.12.027.

Zhou, Y. Y., Li, Y., Jiang, W. Q., & Zhou, L. F. (2015). MAPK/JNK signalling: A potential autophagy regulation pathway. *Bioscience Reports*, *35*(3), e00199. https://doi.org/10.1042/BSR20140141.

Zhou, W., Yue, C., Deng, J., Hu, R., Xu, J., Feng, L., et al. (2013). Autophagic protein Beclin 1 serves as an independent positive prognostic biomarker for non-small cell lung cancer. *PLoS One*, *8*(11), e80338. https://doi.org/10.1371/journal.pone.0080338.

Zhu, H., Wu, H., Liu, X., Li, B., Chen, Y., Ren, X., et al. (2009). Regulation of autophagy by a beclin 1-targeted microRNA, miR-30a, in cancer cells. *Autophagy*, *5*(6), 816–823.

# CHAPTER FIVE

# Recent Advances in Nanoparticle-Based Cancer Drug and Gene Delivery

Narsireddy Amreddy*,†, Anish Babu*,†, Ranganayaki Muralidharan*,†, Janani Panneerselvam*,†, Akhil Srivastava*,†, Rebaz Ahmed*,‡, Meghna Mehta*,†, Anupama Munshi*,†, Rajagopal Ramesh*,†,‡,1

*The University of Oklahoma Health Sciences Center, Oklahoma City, OK, United States
†Stephenson Cancer Center, The University of Oklahoma Health Sciences Center, Oklahoma City, OK, United States
‡Graduate Program in Biomedical Sciences, The University of Oklahoma Health Sciences Center, Oklahoma City, OK, United States
1Corresponding author: e-mail address: rajagopal-ramesh@ouhsc.edu

## Contents

| | | |
|---|---|---|
| 1. | Introduction | 116 |
| 2. | Types of Nanoparticles for Therapeutic Delivery | 117 |
| 3. | Anticancer Therapeutics and Nanoparticle-Based Delivery Agents | 122 |
| | 3.1 Chemotherapeutic Drugs | 123 |
| | 3.2 PDT Drugs | 123 |
| | 3.3 Small Molecule Inhibitors | 126 |
| | 3.4 Nucleic Acid Therapeutic Agents | 127 |
| 4. | Nanoparticle-Based Codelivery of Drugs and Genes | 136 |
| 5. | Stimuli-Responsive Drug Delivery | 139 |
| | 5.1 Thermostimuli | 140 |
| | 5.2 Magnetic Stimuli | 141 |
| | 5.3 Light Stimuli | 142 |
| | 5.4 Ultrasound Stimuli | 144 |
| | 5.5 Redox Stimuli | 145 |
| | 5.6 Enzyme Stimuli | 146 |
| | 5.7 pH Stimuli | 147 |
| 6. | Nanoparticle-Based Receptor-Targeted Delivery | 148 |
| | 6.1 EGFR Receptor | 149 |
| | 6.2 Transferrin Receptor | 151 |
| | 6.3 LHRH Receptor | 152 |
| | 6.4 Folate Receptor | 153 |
| | 6.5 Integrins | 154 |
| 7. | Conclusion and Prospects | 156 |

| | |
|---|---|
| Acknowledgments | 157 |
| References | 157 |
| Further Reading | 170 |

## Abstract

Effective and safe delivery of anticancer agents is among the major challenges in cancer therapy. The majority of anticancer agents are toxic to normal cells, have poor bioavailability, and lack in vivo stability. Recent advancements in nanotechnology provide safe and efficient drug delivery systems for successful delivery of anticancer agents via nanoparticles. The physicochemical and functional properties of the nanoparticle vary for each of these anticancer agents, including chemotherapeutics, nucleic acid-based therapeutics, small molecule inhibitors, and photodynamic agents. The characteristics of the anticancer agents influence the design and development of nanoparticle carriers. This review focuses on strategies of nanoparticle-based drug delivery for various anticancer agents. Recent advancements in the field are also highlighted, with suitable examples from our own research efforts and from the literature.

## 1. INTRODUCTION

Despite advancements in diagnostic and treatment strategies, cancer is still the second most common cause of disease-related death in the United States, according to the American Cancer Society. An estimated 1.6 million new cases of cancer and 0.6 million cancer-related deaths are predicted for the year 2017 (Cancer Facts & Figures, 2017, American Cancer Society). Cancers that spread to many parts of the body require more rigorous and comprehensive treatment regimens that usually involve chemotherapy as the first-line approach (Morabito et al., 2007; Telli & Carlson, 2009; Younes, Pereira, Fares, & Gross, 2011). Most chemotherapy drugs are difficult to administer directly, due to hydrophobicity. In addition, many are toxic to healthy tissues and produce undesirable side effects. As an emerging treatment modality, gene therapy utilizes nucleic acid materials, such as small interfering (si)RNA, DNA, and oligonucleotides, to silence cancer-causing genes, correct mutated genes, or enhance the expression of beneficial proteins that can prevent cancer cell growth and metastasis (Cross & Burmester, 2006; Song, 2007).

Recently, the combined effects of chemotherapy and gene therapy have been realized, especially in overcoming multidrug resistance (MDR). Delivery of gene therapy agents, alone or in combination with chemotherapy, is a hurdle because of their poor stability, lack of tumor selectivity, and rapid clearance from the body (Cross & Burmester, 2006; Wang, Lu, Wientjes, & Au, 2010). Other treatment modalities that use chemical or natural products as anticancer agents or small molecule inhibitors are hampered due to comparable delivery issues. For example, photodynamic therapy (PDT) uses photosensitizers (PSs), which are mostly hydrophobic, lack tumor selectivity, and require soluble formulation for in vivo administration (Babu, Jeyasubramanian, Gunasekaran, & Murugesan, 2012). Therefore, many anticancer agents share similar issues affecting safe and effective delivery to the tumor site via the systemic route.

To overcome these limitations, various delivery vehicles have been developed to achieve high therapeutic efficiency of anticancer agents by providing protection in the circulation and enhancing their bioavailability (Haley & Frenkel, 2008). Nanoparticles are attractive vehicles for anticancer agents, because of their controlled drug release and tumor-selective properties (Prabhu, Uzzaman, & Guruvayoorappan, 2011). Nanoparticles have been developed based on liposomes/lipids, polymers of synthetic and natural origin, and inorganic particles (Panyam & Labhasetwar, 2003; Wang, Li, Cheng, & Yuan, 2016; Zhao & Huang, 2014). This review describes the strategies using nanoparticles to deliver anticancer agents, such as chemotherapy drugs, PDT agents, small molecule inhibitors, and therapeutic genes. Nanoparticle designs for stimuli-responsive drug release are also emphasized. In addition, this review outlines some of the major cancer-specific cell surface receptors that have been explored for ligand-based targeted drug delivery using nanoparticles.

## 2. TYPES OF NANOPARTICLES FOR THERAPEUTIC DELIVERY

Liposome-based nanoparticle systems are widely used for drug and gene delivery. Since the 1995 US Food and Drug Administration (FDA) approval of the liposomal formulation for anticancer chemotherapeutic doxorubicin, Doxil®, many new liposomal formulations have been developed, entered into clinical trials, and FDA approved for clinical use (Table 1).

Liposomes are small nanosized vesicles formed by lipid bilayers with an aqueous core. Hydrophilic drugs are encapsulated inside the core, whereas hydrophobic drugs are entrapped in the lipid bilayer. If the active agent is nucleic acid, liposomes are cationic in nature (Simoes et al., 2005). These cationic liposomes form a complex with negatively charged nucleic acid

Table 1 Clinically Approved Liposomal Chemotherapeutic Formulations for Cancer Therapy

| Brand Name | Drug (Active Agent) | Liposome Composition | Target Cancer | Manufacturer |
|---|---|---|---|---|
| Doxil | Doxorubicin | HSPC:Cholesterol:PEG 2000–DSPE | Ovarian, Breast, AIDS-Related Kaposi's Sarcoma | Ben Venue Laboratories |
| Lipodox | Doxorubicin | HSPC:Cholesterol:PEG 2000–DSPE | Ovarian, Breast, AIDS-Related Kaposi's Sarcoma, Multiple Myeloma | Sun Pharmaceuticals |
| Myocet | Doxorubicin | EPC:Cholesterol | Breast | Teva Pharma B.V. |
| DaunoXome | Daunorubicin | DSPC:Cholesterol | AIDS-Related Kaposi's Sarcoma | NeXtar Pharmaceuticals |
| Mepact | Mifamurtide | DOPS:POPC | Osteosarcoma | Takeda Pharmaceuticals Limited |
| Onivyde | Irinotecan | DSPC:MPEG-2000:DSPE | Pancreatic Adenocarcinoma | Merrimack Pharmaceuticals Limited |
| Marqibo | Vincristine | Sphingomyelin and Cholesterol | Philadelphia Chromosome–Negative (Ph−) Acute Lymphoblastic Leukemia | Talon Therapeutics, Inc. |
| Lipoplatin/ Nanoplatin | Cisplatin | Soy Phosphatidyl Choline (SPC-3): Cholesterol mPEG-DSPE:DPPG | Pancreatic, Lung | Regulon, Inc. |

therapeutics through electrostatic interaction. If the liposomes are anionic, the nucleic acid materials are either complexed in the presence of calcium ions or encapsulated inside the core (Patil, Rhodes, & Burgess, 2004). Neutral lipids are used as helper lipids for efficient gene delivery (Balazs & Godbey, 2011). The major limitation of liposomes is their poor stability and burst release of drug.

Solid lipid nanoparticles (SLNs) offer unique properties, including small size, large surface area, high drug loading, the interaction of phases at the interfaces, and improved therapeutic efficacy of the loaded drug. SLNs are also more stable than liposomes and offer controlled release of their cargo (Muller, Mader, & Gohla, 2000). SLNs consist of solid-phase lipids at room temperature and surfactants for emulsification.

Polymer nanoparticles form a significant category of drug delivery vehicles, in which both synthetic and natural polymers are used in the preparation of nanoparticles (Amreddy, Babu, Muralidharan, Munshi, & Ramesh, 2017). Polymer nanoparticles are versatile in delivering numerous materials, including chemotherapy drugs, small molecules, genes, and proteins. Numerous polymer nanoparticles, such as polylactic acid (PLA), polylactic acid-*co*-glycolic acid (PLGA), poly(alkyl cyanoacrylate) (PACA), polycaprolactone (PCL), polyanhydrides, polyethyleneimine (PEI), chitosan, and gelatin, are being tested in the laboratory (Zhang & Saltzman, 2013). The aforementioned polymers are biodegradable. Among these polymers, PLGA has been approved by the FDA for drug delivery purposes. Lupron Depot® is an FDA-approved drug for prostate cancer therapy that is based on a PLGA nanoparticle platform (Zhang & Saltzman, 2013). PCL nanoparticles are another controlled release platform for chemotherapeutics and other anticancer agents. They are commonly used as a long-term controlled delivery system because of their slow degradation rate (Danafar & Schumacher, 2016). In addition, polyanhydride-controlled release systems can be manipulated to release drugs during a large time range by adjusting the ratios of the copolymers' aliphatic and aromatic anhydrides (Fu & Kao, 2010). Such a polyanhydride system, GLIADEL®, is used in the treatment of recurrent malignant glioma (Perry, Chambers, Spithoff, & Laperriere, 2007). Furthermore, PEI is a cationic linear or branched polymer commonly used for the delivery of siRNA or DNA for gene therapy (Lungwitz, Breunig, Blunk, & Gopferich, 2005). PEI has numerous free cationic groups that can electrostatically interact with nucleic acids and thereby condense them to form nanosized particles. However, PEI has a major limitation in that it is toxic to cells. Modification of PEI with neutral polymer PEG

(Wu et al., 2010) or hybridization with other biocompatible polymer could reduce its toxicity and enhance its in vivo stability (Zheng et al., 2012).

Chitosan and gelatin are natural polymers that are commonly used for drug or gene delivery. Chitosan is a cationic polysaccharide, a modified form of chitin that is known for its biocompatibility and biodegradability. The presence of numerous functional groups allows for easy modification for use in drug delivery applications. Chitosan nanoparticles can encapsulate chemotherapeutics, genes, peptides, or small molecules, and efficiently deliver the cargo for tumor therapy (Babu & Ramesh, 2017; Layek, Lipp, & Singh, 2015; Patel, Patel, & Patel, 2010). The cationic nature of chitosan is harnessed for gene delivery, as negatively charged nucleic acids readily form complexes with chitosan to form nanoparticles. Gelatin-based nanocarriers are also highly biocompatible and biodegradable and can be easily functionalized due to their numerous free chemical groups. Gelatin is often crosslinked to form stable nanoparticles and to improve its performance as a drug delivery system. Gelatin nanocarriers can deliver a variety of drugs, including chemotherapeutics, protein and peptides, and siRNA (Kulsharova et al., 2013; Lee, Yhee, Kim, Kwon, & Kim, 2013; Xu, Singh, & Amiji, 2014).

Other commonly used nanocarriers include dendrimers, metal-based nanoparticles, such as gold and iron oxide nanoparticles, quantum dots, cyclodextrin, and silica nanoparticles. Dendrimers form unique nanocarriers and are uniformly branched structures originating from a core molecule. Numerous multivalent surface functional groups offer wide potential for multiple interactions and modifications, and can link therapeutic molecules and nucleic acids onto the surface functional groups and in the cavities formed by the branches radiating from the core molecule (Dufes, Uchegbu, & Schatzlein, 2005; Madaan, Kumar, Poonia, Lather, & Pandita, 2014).

The most commonly used dendrimer for drug and gene delivery applications is the polyamidoamine (PAMAM) dendrimer. PAMAM dendrimers of different generations can have drugs loaded into their cavities or attached to the periphery using chemical methods, and controlled and specified drug delivery can occur. Moreover, PAMAM dendrimers can also carry nucleic acid therapeutics, aided by electrostatic interaction-based complex formation (Abbasi et al., 2014). Most of these characteristics make dendrimers a good choice for drug and gene delivery for cancer therapy.

Gold nanoparticles exhibit a combination of unique physicochemical and optical properties that permit their use in various biomedical applications, such as diagnostic imaging and drug and gene delivery (Cai, Gao, Hong, & Sun, 2008). Gold nanoparticles are typically very small (<10 nm), can easily penetrate into cells, and have the capability to deliver drugs, genes, and imaging agents with low solubility and poor pharmacokinetics. In general, gold nanoparticles exist in different shapes and structures, such as spheres, rods, stars, and clusters. All of these shapes have been explored for drug delivery. Gold nanoparticles are routinely surface functionalized with ligands to achieve increased selectivity in tumors and to specifically deliver their payload. These nanoparticles are generally modified with polymers or PEG-containing linkers to conjugate or complex with drug or siRNA/DNA. Hence, they are increasingly being used for drug or gene delivery purposes by exploiting the advantages of increased drug/gene loading, low toxicity due to the inert nature of gold, efficiency in cell uptake, fast endosomal escape, and stability in the circulation (Mendes, Fernandes, & Baptista, 2017).

Iron oxide nanoparticles, otherwise known as magnetic nanoparticles, have been used in biomedical applications for the last 3 decades. However, the potential of magnetic nanoparticles in drug and gene delivery for cancer therapy has only recently been realized. Therapeutic agents are either attached to or encapsulated within the nanoparticle. These particles have either magnetic cores with a polymer or metal coating that can be modified, or a porous polymer matrix that has magnetic particles precipitated within the pores (McBain, Yiu, & Dobson, 2008). Under the influence of an external magnetic field, magnetic nanoparticles are attracted to the tumor site and deliver the drugs. This is an important advantage, because this strategy can prevent accumulation of drug in healthy tissue. Superparamagnetic nanoparticles behave like magnets only when this external magnetic field is applied, and cause no toxicity on their own.

Mesoporous silica nanoparticles (MSNs) are a special class of nanoparticles that are small, rigid, and nonbiodegradable nontoxic nanocarriers for drugs, peptides, and genes. The characteristics of MSNs include a mesoporous nature, tunable pore size, high drug loading efficiency, and surface properties that can be altered to favor efficient drug or gene delivery, depending on additives used in the preparation of MSNs (Bharti, Nagaich, Pal, & Gulati, 2015). To enhance the surface properties, PEG molecules can be used to modify the MSNs surface by covalent conjugation or by adsorption or entrapment of PEG molecules in the MSNs surface. PEG functionalization can be utilized for covalent conjugation of drug molecules and targeting ligands for selective drug delivery. For gene delivery, a cationic polymer, such as PEI, is useful in modifying the MSNs surface to electrostatically attract nuclei acid molecules for delivery. The presence of PEI also enhances the endolysosomal escape capability of MSNs carrying nucleic acids (Hom et al., 2010).

Recently, researchers have explored the possibilities of using extracellular vesicles as drug/gene delivery vehicles (Jiang, Vader, & Schiffelers, 2017). Exosomes are the most prominently studied (Jiang et al., 2017; Srivastava, Amreddy, et al., 2016; Srivastava, Babu, et al., 2016). Exosomes are nanosized (30–100 nm) lipid bilayer cellular vesicles that are involved in transportation of cellular cargo. Because of their small size, cellular origin, flexibility to incorporate macromolecules, such as DNA, RNA, and micro-RNAs (miRNA), into their lumen, and ability to cross-stringent biological barriers, such as the blood–brain barrier, exosomes are considered ideal candidates for drug/gene delivery vehicles (Shahabipour et al., 2017). Studies have shown that therapeutic siRNA and miRNA can be inserted into exosomes either by transforming the cell producing exosomes or by directly manipulating their lumen by adding molecules using techniques such as incubation, sonication, or electroporation (Johnsen et al., 2014). Although the field of exosome biology and application is relatively new, it is predicted that exosomes will have an important role in personalized and precision medicine, as patients' own cells can be used to produce exosomes that will be used as therapeutic carriers.

Thus, a multitude of nanoparticle formulations are currently under investigation in the laboratory and clinical settings to enhance drug delivery efficacy for cancer therapy. Therefore, it is expected that the use of nanoparticle drug delivery systems will spread rapidly across different human ailments, especially cancer. The following sections discuss the different types of therapeutic molecules and delivery strategies using specifically designed nanoparticles, citing recent examples.

## 3. ANTICANCER THERAPEUTICS AND NANOPARTICLE-BASED DELIVERY AGENTS

Anticancer agents such as chemotherapeutics, photodynamic agents, small molecule inhibitors, and nucleic acid-based therapeutic agents often requires safe and efficient delivery systems for stability enhancement, delaying fast bioclearance, high accumulation in tumor site, and enhanced therapeutic potential upon systemic administration. In addition, nanoparticle-based drug delivery systems offer minimized exposure of drug toward normal tissues in the body. This section discusses different kinds of anticancer agents and nanoparticle-based drug delivery strategies for cancer therapy with suitable examples.

## 3.1 Chemotherapeutic Drugs

In cancer treatment, chemotherapy has been practiced for more than a century. Numerous antitumor agents that inhibit and reduce tumor growth are available (Isoldi, Visconti, & Castrucci, 2005; Monks et al., 1991). Chemotherapeutic agents include alkylating agents, plant alkaloids, antitumor antibiotics, and antimetabolics. Some agents are specifically used against certain cancers, while others are used in the treatment of multiple cancers. For example, hydrophilic drugs, such as doxorubicin (DOX), platinum compounds (cisplatin, carboplatin, and oxaliplatin), 5-fluorouracil, gefitinib, epirubicin, paclitaxel, camptothecin, and curcumin are used to treat various cancers. Most chemotherapeutic agents target cell division, eventually inhibiting cell growth via mechanisms such as inhibiting microtubule function, protein function, or DNA synthesis (Dumontet & Sikic, 1999; Palchaudhuri & Hergenrother, 2007). The direct administration of drugs into the body can cause severe side effects and toxicity to normal healthy tissues (Celikoglu, Karayel, Demirci, Celikoglu, & Cagatay, 1997). Hence, chemotherapy requires a withdrawal period to allow patients to recover from the effects of toxicity. Therefore, researchers are focused on the development of systems to deliver antitumor drugs to the targeted location. The specific delivery of drugs can be achieved safely by optimizing drug dose and monitoring drug levels in the body. These methods must allow for a controlled rate of delivery, sustained release, and targeted delivery to improve therapeutic efficacy while simultaneously reducing toxicity to normal tissues.

## 3.2 PDT Drugs

PDT is an emerging approach for improved and effective cancer treatment. Compared with other cancer treatment approaches, PDT has several advantages. PDT is cost effective and reduces the need for repeated and prolonged treatment. Most PSs exhibit auto fluorescence, which makes them easy to track and observe uptake inside the body (Usuda et al., 2007). Moreover, PDT can be controlled with near-infrared (NIR) light irradiation; once in the tumor region of interest, the PS can be irradiated and the drug released, thus reducing toxicity to normal tissues

(Wang, Tao, Cheng, & Liu, 2011). PDT is mainly recommended for oral, head and neck, and ocular cancers, although researchers are currently exploring the use of PDT in the treatment of other cancers (Yoshida et al., 2008). PDT occurs through the dynamic nature of the three essential components: NIR light, PS (e.g., porphyrins, phthalocyanines, and chlorines), and molecular oxygen. The PS generates reactive oxygen species (ROS) in the presence of light, and causes oxidative damage to DNA and proteins that are involved in cell growth and proliferation of the tumor and its surrounding vasculature, as illustrated in Fig. 1 (Castano, Demidova, & Hamblin, 2004). The physical and chemical properties of the materials that are used in PDT will influence the effectiveness of the therapy. Injecting the PS directly into the body results in nonspecific biodistribution and poor tumor accumulation. Most PSs are hydrophobic in nature. Thus, the hydrophobic PS exhibits poor solubility under physiological conditions that reduce the tumor-targeting ability. The specific delivery of PS into tumor tissue is a challenge and requires a carrier that specifically delivers the PS into the targeted tissue region. Nanoparticles are appropriate carriers for delivering PS molecules. By conjugating or encapsulating the PS in nanoparticles, the above-mentioned limitations could be addressed (Chatterjee, Fong, & Zhang, 2008). In this section, we discuss recent advances in PDT with nanoparticles.

Flak et al. demonstrated that ZnPc@TiO$_2$ hybrid nanostructures, in the form of nanoparticles and nanotubes, could be used for dual therapy by loading PDT agent zinc phthalocyanine (ZnPc) and anticancer drug doxorubicin with folic acid targeting. They observed that these hybrid

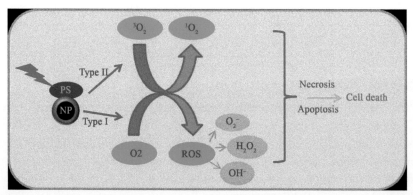

**Fig. 1** Schematic representation of the types I and II mechanisms of PDT upon light irradiation to generate ROS and singlet oxygen that cause necrosis and apoptosis, respectively.

nanoparticles selectively targeted and displayed more uptake in human cervical cancer cells (HeLa) than in normal fibroblasts (MSU-1.1). Enhanced in vitro cytotoxicity and photocytotoxic activity was demonstrated with hybrid nanostructures selectively targeting to cancer cells (Flak, Yate, Nowaczyk, & Jurga, 2017).

Recently, Wang et al. used cold atmospheric plasma (CAP) as a light source, instead of NIR light. The researchers used Protoporphyrin IX (PpIX), which is the most popular PDT PS, using polymerosomes as carrier. They demonstrated that cancer cells can be effectively killed by using a CAP light source with a wide range of wavelengths, compared with UV light sources. After CAP irradiation, melanoma cell killing efficiency significantly increased to 80% when compared with no light source (45%) or a UV light source (65%) (Wang, Geilich, Keidar, & Webster, 2017).

Gold nanoparticles can be used as photothermal therapy (PTT) agents that show a synergistic therapeutic effect when combined with PDT. Li and colleagues reported the use of gold nanoclusters for dual PTT/PDT therapy for pancreatic ductal adenocarcinoma (PDAC). Gold nanoclusters were used as the PTT agent and Cathepsin E (CTSE) was used as the PDT prodrug. In this system, they observed that CTSE-sensitive nanoclusters with targeting U11 peptide can significantly increase the uptake and apoptosis of pancreatic cancer cells, compared with that of the nontargeted nanocluster AuS-PEG and the insensitive nanocluster AuC-PEG. They reported that enzyme-triggered drug release of 5-ALA with tumor targeting in nanoclusters AuS-U11 could achieve optimal therapeutic efficacy with endomicroscopy-guided PTT/PDT, with reduced side effects (Li, Wang, et al., 2017).

Gold nanorods are used as PTT agents with high laser doses, usually 808 nm, 1–48 W/cm$^2$ irradiations that kill the cancer cells. Vankayala et al. reported that gold nanorods themselves act as PDT and PTT therapeutic agents by optimizing the irradiation wavelengths. The researchers demonstrated that gold nanorods alone can sensitize the formation of singlet oxygen ($^1O_2$) as PDT affects inhibition of tumors in mice under very low LED/laser doses of single photon NIR (915 nm, <130 mW/cm$^2$) light excitation. By changing the NIR light excitation wavelengths, Au NRs-mediated phototherapeutic effects can be switched from PDT to PTT or a combination (Vankayala, Huang, Kalluru, Chiang, & Hwang, 2014).

Magnetic nanoparticles are well known as magnetic resonance (MR) imaging agents. When a PDT agent is delivered with magnetic nanoparticles, the magnetic nanoparticles are used for image-guided PDT cancer therapy, which integrates diagnostic and therapeutic functionalities into a single system. Zhou et al. conjugated the IR820 onto the surface of iron oxide nanoparticles with 6-amino hexanoic acid to form IR820-CSQ-Fe conjugates. The IR820-conjugated iron oxide nanoparticles showed an enhanced ability to produce singlet oxygen almost double that of free dyes, which improved its efficiency for PDT, and showed increased T1 and T2 relaxation values (Zhou et al., 2016). In another approach, magnetic and fluorescent lanthanide-doped gadolinium oxide ($Gd_2O_3$) UCNs with bright upconversion luminescence (UCL) and high longitudinal relaxivity ($r_1$) were used for simultaneous magnetic resonance imaging (MRI)/UCL dual-modal imaging and PDT. The results showed that these UCN/UCL/MRI nanoparticles act as good MRI contrast agents and fluorescence nanoprobes for live cell imaging. The researchers utilized the luminescence-emission capability of the UCNs for the activation of a PS to achieve significant PDT results and image-guided delivery (Liu, Huang, et al., 2017).

## 3.3 Small Molecule Inhibitors

Small molecule inhibitors are organic compounds with different chemical structures and molecular weights ranging from 500 to 900 Da; these inhibitors can easily diffuse across cell membranes to reach intracellular sites (Veber et al., 2002). Once the inhibitor enters the cells, it affects various other molecules, such as proteins, and ultimately kills cancer cells (Arkin & Wells, 2004). Small molecule inhibitors can be classified as natural, such as secondary metabolites, or artificial, such as antiviral drugs. These agents are structurally similar to chemotherapy drugs, but function as RNAi to knockdown specific proteins. Small molecule inhibitors can be used to study aspects of cell biology, such as cell cycle control, mitosis, oncogenic signaling pathways, gene expression, apoptosis, and autophagy (Weiss, Taylor, & Shokat, 2007). Inhibitors are used either alone or encapsulated with nanoparticles to treat various cancers.

Small molecule inhibitors have advantages over larger molecules, such as monoclonal antibodies. These advantages include easy diffusion across the cell membrane, which facilitates large-scale experiments in which there might be difficulties in transfection or knockdown using RNAi (Gowda, Jones, Banerjee, & Robertson, 2013; Madani, Naderi, Dissanayake, Tan, & Seifalian, 2011; Parhi, Mohanty, & Sahoo, 2012; Perumal, Banerjee, Das, Sen, & Mandal, 2011). Small molecule inhibitors can be administered orally, whereas biologics generally require injection or another parenteral administration. These inhibitors can be easily combined with other treatments. Titration of doses is relatively easy, and the compounds can be used for different cell lines and even from different species.

MiaPaCa-2 tumor cells treated with PH-427, a small molecule inhibitor of AKT/PDK1, showed rapid internalization. However, PH-427 encapsulated with PNP (PH-427-PNP) treatment improved the therapeutic effect in in vitro and in vivo studies in pancreatic cancer cells (Lucero-Acuna et al., 2014). Encapsulation of Mcl-1 small molecule inhibitors with lipid nanoparticles showed better cell killing efficiency in monocytes infected with human cytomegalovirus (HCMV) (Burrer et al., 2017).

Prostate-specific membrane antigen (PSMA) is a diagnostic and therapeutic marker for prostate cancer. Targeted delivery of gold nanoparticles with small molecule inhibitor phosphoramidate peptidomimetic to PSMA-expressing prostate cancer cells showed significantly higher and selective binding to LNCap cells than did nontargeted gold nanoparticles (Kasten, Liu, Nedrow-Byers, Benny, & Berkman, 2013). The small molecule inhibitor for Niemann–Pick type C1 (NPC1), NP3.47, formulated with lipid nanoparticles containing siRNA, enhanced the therapeutic efficacy of LNP–siRNA by increasing endosomal escape (Wang, Tam, et al., 2016).

Polymeric biodegradable nanoparticle poly-L-glutamic acid (PGA) is used for the conjugation of small molecules. Xyotax (PGA-paclitaxel) and CT-2106 (PGA-camptothecin) are now in clinical trials (Bhatt et al., 2003; Sabbatini et al., 2004). Hh pathway inhibitor (HPI)-1 is an antagonist of Gli1. It has been shown that encapsulation of HPI-1 using PLGA conjugated with PEG improved the systemic bioavailability compared with the parent compound in pancreatic cancer (Chenna et al., 2012). Aurora B is involved in cytokinesis, and inhibition of this kinase has shown mitotic catastrophe. Encapsulation of AZD2811 and AZD1152 using polymeric and lipid nanoparticles has resulted in increased drug accumulation in tumors (Ashton et al., 2016).

## 3.4 Nucleic Acid Therapeutic Agents

Targeted regulation of the expression of genes that are known to be involved in cancer-related pathways is the ultimate goal of various experimental approaches for cancer treatment. Efforts are being made to achieve complete knockdown of the targeted gene to stop the downstream network of genes/proteins, preventing disease progression. The initial efforts to knockdown genes involved the antisense method, ribozymes, and chimeric oligonucleotides. However, only a very weak suppression of gene expression was achieved (Watts & Corey, 2012).

In the 1990s, a novel method of controlling gene expression was discovered in a worm model organism (*Caenorhabditis elegans*). A short double-stranded RNA (dsRNA) was used to specifically and selectively suppress the expression of the target gene (Fire et al., 1998). This phenomenon is called RNAi or RNA interference. The dsRNA was termed small interfering RNA, or siRNA. RNAi begins when an enzyme, DICER, encounters dsRNA and chops it into pieces called small interfering RNAs or siRNAs (Bernstein et al., 2003). This enzyme belongs to the RNase III nuclease family. A complex of proteins gathers up these RNA remains and uses their code as a guide to search out and destroy any RNAs with a matching sequence, such as target mRNA, in the cell. The mechanism of RNAi-mediated gene silencing is commonly called posttranscription gene silencing, or PTGS (Zamore, Tuschl, Sharp, & Bartel, 2000).

In cancer treatment, RNAi technology has opened a new arena of opportunities to develop innovative gene therapy strategies. The targeted approach and the ability to design and custom synthesize siRNA with the desired sequence can be exploited to selectively knockdown the expression of specific genes that are critical for pathophysiology of cancer, such as different signaling pathways involved in cancer cell growth, metastasis, angiogenesis, or drug resistance. In the following sections, we discuss the advancements and methods for using siRNA, DNA, shRNA, and miRNA with different types of nanocarriers.

### 3.4.1 siRNA Delivery

Although the phenomenon of siRNA-based cancer gene therapy is promising, direct administration of siRNA to cells did not yield the expected results, largely because naked siRNA has a short half-life in vivo, as it is rapidly degraded by nucleases in biological fluids (Choung, Kim, Kim, Park, & Choi, 2006). Furthermore, with an average size of 15 kDa, the naked siRNA cannot penetrate endothelial cells and tissues, and the extracellular matrix (ECM) further prevents siRNA from diffusing efficiently into the cellular membrane. In addition to these biological barriers, the negative charge present on siRNA prevents it from penetrating through negatively charged cell membranes. Inside the cell, the siRNA encounters harsh environment produced by the endosomes, which often results in clearing of siRNA through endosomal pathways. Finally, siRNA is cleared by the site of disease and bioaccumulates in the kidneys and liver. To circumvent these limitations and exploit the benefits of PTGS of targeted cancer-related genes by siRNA, researchers have developed novel nanoparticle-based delivery vehicles to transport siRNA into cells at the site of action in cancer cells. This method of delivery has additional benefits, such as that target moieties can be added to the nanoparticles that cause targeted delivery of complex to the desired cells. However, the physicochemical properties must be considered before selecting a delivery vehicle for siRNAs, as described in Tatiparti, Sau, Kashaw, and Iyer (2017).

Several nanoparticle formulations combining lipids and polymers have been developed as hybrid delivery systems that harness the properties of lipids, which facilitate the penetration and efficient delivery of siRNA to the cell, and properties of polymers, which provide stability and biocompatibility. Some of the siRNA delivery systems using liposome and polymer nanoparticles are in clinical trials, as represented in Table 2. A similar hybrid delivery system delivery of p53 siRNA was successfully tested by Kundu et al. in a mouse osteocarcinoma cell line (Kundu et al., 2017). Neutral lipids are biocompatible, but have fewer interactions with anionic polynucleotides, resulting in their poor bioavailability. The siRNA can be loaded inside the core of nanoparticles or can be coated into the surface of nanoparticles. Organic in nature, these nanoparticles show low immunogenic response and are biodegradable. Furthermore, siRNA can be complexed with lipids for additional stability of the siRNA in complexes, which promotes endocytosis of siRNA into the cells, as shown by Kim et al. (2007) in their work combining siRNA with DOTAP/cholesterol. In ovarian cancer liposomes, zwitterion 1,2-dioleoyl-sn-glycero-3-phosphatidylcholine (DOPC) was added to provide optimum stability and was shown to efficiently target oncoprotein EphA2 (Landen et al., 2005).

Chitosan is another negatively charged biopolymer that has been explored for drug/gene delivery formulations because of physical compatibility. Anderson et al. reported delivery of siRNA in lung adenocarcinoma H1299 cells via cationic chitosan (Andersen, Howard, Paludan, Besenbacher, & Kjems, 2008). The formulations of chitosan/siRNA nanoparticles vary, especially in the ratio of molecular weight to charge, which affects its stability and efficiency (Hsu et al., 2013).

Metal-based nanoparticles, such as gold or iron oxide and carbon materials, are also used to deliver siRNA. Wang et al. used an iron oxide-based nanoparticle vector for tumor-targeted siRNA delivery in an orthotropic

Table 2 Current Clinical Trials With Different siRNA Delivery Platforms

| NCT Number | Title | Recruitment | Conditions | Interventions | Gender | Phases | Mode of Delivery |
|---|---|---|---|---|---|---|---|
| NCT00938574 | Study With Atu027 in Advanced Solid Cancer | Completed | Advanced Solid Tumors | Drug: Atu027 | All | Phase 1 | Lipid nanoparticle |
| NCT00689065 | Safety Study of CALAA-01 in Solid Tumor Cancers | Terminated | Cancer Solid Tumor | Drug: CALAA-01 | All | Phase 1 | Polymer nanoparticle |
| NCT02166255 | APN401 in Treating Cancer That Are Metastatic or Cannot Be Removed By Surgery | Active, not recruiting | Recurrent tumors at different stages | Biological: siRNA | All | Phase 1 | siRNA transfection |
| NCT01591356 | EphA2 Gene Targeting | Recruiting | Advanced Cancers | Drug: siRNA-EphA2-DOPC | All | Phase 1 | Lipid nanoparticle |
| NCT01437007 | TKM 080301 for Primary or Secondary Liver Cancer | Completed | Cancers With Hepatic Metastases | Drug: TKM-080301 | All | Phase 1 | Lipid nanoparticle |
| NCT02110563 | Study of DCR–MYC in Solid Tumors, Multiple Myeloma, or Lymphoma | Active, not recruiting | Solid Tumors | Drug: DCR-MYC | All | Phase 1 | Lipid nanoparticle |

| NCT01808638 | Atu027 Plus Gemcitabine in Advanced or Metastatic Pancreatic Cancer | Completed | Carcinoma, Pancreatic Ductal | Drug: Atu027 and gemcitabine | All | Phase 1\|Phase 2 | Lipid nanoparticle |
|---|---|---|---|---|---|---|---|
| NCT00672542 | Immunotherapy of Melanoma With siRNA | Completed | Metastatic Melanoma\|Absence of CNS Metastases | siRNA | All | Phase 1 | Transfection |
| NCT01676259 | Study of siG12D LODER in Combination With Chemotherapy | Not yet recruiting | Pancreatic Cancer | Drug: siG12D-LODER\|Drug: Gemcitabine + nab-Paclitaxel | All | Phase 2 | Polymer nanoparticle |
| NCT02314052 | DCR-MYC in Hepatocellular Carcinoma | Active, not recruiting | Hepatocellular Carcinoma | Drug: DCR-MYC | All | Phase 1\|Phase 2 | |

Data retrieved from www.clinicaltrial.gov on June 28, 2017.

hepatocellular carcinoma xenograft mouse model (Wang, Kievit, et al., 2016). Hybrid nanoparticles are the latest approach to explore improvements in gene delivery efficiency by overcoming the limitations associated with individual methods. Sardo et al. reported that transfection efficiency and low bioavailability can be considerably improved by using polymer-modified gold nanostars. They coated the gold nanostar with layers of hydrophilic and amphiphilic polymers that contained –SH and –SS linker groups to anchor the siRNA layers. The results indicated improved transfection efficiency and enhanced colloidal stability (Sardo et al., 2017). In another example, Lee et al. fabricated a hybrid delivery system by combining chitosan–deoxycholic acid with perflouropentane and iron oxide. To this hybrid system, siRNA was electrostatistically loaded and delivered to lung and breast cancer cell lines. The results demonstrated enhanced delivery and effect of siRNA in the recipient cells upon ultrasound exposure (Lee et al., 2017).

Since cancer is a multifaceted disease, the pooling of multiple siRNA has been considered. Chen et al. described a strategy for using a metallic nanoparticle consisting of selenium and ruthenium to deliver pools of siRNA for MDR and to interfere with microtubule dynamics. They showed enhanced uptake of siRNA in breast cancer cells and induction of cytotoxic and downstream signaling pathways, leading to improved therapeutic effects (Chen, Xu, et al., 2017).

Iron oxide nanoparticles can be used during MRI. Hence, delivery of therapeutic siRNA using these nanoparticles can be used for the dual purposes of noninvasive cancer imaging and treatment (Medarova, Balcioglu, & Yigit, 2016). Many recent studies have demonstrated the flexibility of gold-, iron-, and silver-based metallic nanoparticles when developing hybrid nanoparticles. This flexibility offers increasing possibilities to further develop metallic nanoparticles that could be used for delivery of siRNA in clinics in the near future.

siRNA efficacy would be further improved if the cells' own vesicles could be used as delivery vehicles. This approach will overcome the body's immunogenic response. Exosomes, the 30–100-nm-sized nanovesicles produced from almost all cells, have attracted recent attention as promising drug delivery vehicles. Previously considered the trash bag of the cell, with the major function of removing toxic materials from the cell, exosomes are now recognized for transporting cellular molecules, such as nucleic acids, proteins, and lipids, from one cell to another and/or to intercellular spaces. Alvarez-Ervit et al. used exosomes derived from immature dendritic cells to deliver siRNA for glyceraldehyde 3-phosphate dehydrogenase (GAPDH) and BACE1 into the brains of C57BL/6 mice, demonstrating for the first time that siRNA can be delivered via exosomes. Exosome-based vehicles are reported to overcome cytotoxic and immunogenic responses produced by synthetic nanoparticles. Since exosomes can pass through the stringent blood–brain barrier, exosome-based siRNA delivery for genes involved in brain cancer could be successful in the future (Alvarez-Erviti et al., 2011).

### 3.4.2 Plasmid DNA Delivery

DNAs are high molecular-weight molecules with double strands that contain transgenes for specific proteins. In therapeutics, DNA inhibits the generation of specific protein through the release of transgenes into cells. The design and construction of plasmids with transgene(s) of interest is an important step in developing DNA therapeutics. Nanoparticle-based carriers play an important role in improving transfection efficiency and therapeutic effects. Cationic DOTAP liposome formulations are used for the conjugation of CXCR4 siRNA, HMGA1 DNA, IL-24 DNA, and FUS1 DNA in lung adenocarcinoma cells (Ito et al., 2004; Panneerselvam et al., 2015, 2016). Codelivery of p53-encoding plasmid and DOX using double-walled PLGA microspheres reportedly improved drug activity (Xu, Xia, Wang, & Pack, 2012). A nonviral dendrimer-based delivery system was used to deliver a combination of TRAIL gene therapy with DOX chemotherapy in human glioblastoma (Liu et al., 2012). Furthermore, SLNs composed of 3β-[$N$-($N'$,$N'$-dimethylaminoethane)-carbamoyl]cholesterol hydrochloride (DC-Chol), 1,2-dioleoyl-sn-glycero-3-phosphoethanolamine (DOPE), and Tween 80 were used to complex with p53-EGFP plasmid DNA in NSCLC cells (Choi et al., 2008). Lecithin-based SNLs were used for the combination therapy of EGFP plasmid DNA and doxorubicin in mice bearing A549 tumors (Han, Zhang, Chen, Sun, & Kong, 2014). Cationic poly-amino acid conjugated magnetic nanoparticles were complexed with NM230HA-GFP plasmid DNA for gene transfection in mice bearing B16F10 melanoma tumors (Jiang, Eltoukhy, Love, Langer, & Anderson, 2013). Nonviral DNA delivery using biodegradable poly(β-amino ester)s (PBAEs) has been shown to be more effective than naked DNA in targeting malignant glioma (Guerrero-Cazares et al., 2014).

### 3.4.3 shRNA Delivery

shRNA, or short hairpin or small hairpin RNA, is another mode of inducing RNA interference-mediated posttranscriptional gene silencing for target genes. The shRNA consists of an RNA molecule with a hairpin-like structure; the molecule is slightly larger than siRNA molecules and, unlike siRNA, is produced inside the cell in the nucleus. From a therapeutics perspective, shRNA can be used to treat different diseases, including cancers. Clinical trials using shRNA are listed in Table 3. Several shRNA delivery methods have been studied. In prostate cancer, a folate-targeted nanoparticle complexed with AR-shRNA enhanced radiosensitivity (Zhang, Liu, et al., 2017). In addition, WT-shRNA was delivered using a transferrin-targeted PEG liposome in melanoma (Saavedra-Alonso et al., 2016).

### 3.4.4 miRNA Delivery

miRNAs are another class of regulatory RNA epigenetically controlling various cellular processes under normal and pathological conditions. These miRNAs are conventionally classified as a small, noncoding class of RNA with a size of 18–22 nucleotides. The mechanism of action of miRNA

**Table 3** Current Clinical Trials With Different shRNA Delivery Platforms

| Title | Recruitment | Conditions | Interventions |
|---|---|---|---|
| Bi-shRNA-furin and Granulocyte Macrophage Colony Stimulating Factor Tumor Cell Vaccine for Advanced Cancer | Active, not recruiting | Ewing's Sarcoma, Nonsmall Cell Lung Cancer, Liver Cancer | Biological: Vigil™\|Biological: Vigil™\|Biological: Vigil™ |
| Pbi-shRNA STMN1 LP in Advanced and/or Metastatic Cancer | Completed | Advanced Cancer Metastatic Cancer Solid Tumors | Biological: pbi-shRNA STMN1 LP |
| Pbi-shRNA™ EWS/FLI1 Type 1 LPX in Advanced Ewing's Sarcoma | Recruiting | Ewing's Sarcoma | Biological: pbi-shRNA™ EWS/FLI1 Type 1 LPX |
| Treatment of Chronic Lymphocytic Leukemia | Not yet recruiting | Leukemia, Lymphocytic, Chronic, B-cell | Genetic: shRNA |
| Human Immunodeficiency Virus-Related Lymphoma Receiving Stem Cell Transplant | Recruiting | HIV Infection, Mature T cell and NK cell, and cancer | Lentivirus Vector CCR5 shRNA and other interventions |

Data retrieved from www.clinicaltrial.gov on June 28, 2017.

overlaps with that of siRNA in certain aspects, but miRNA can regulate more than one mRNA and, thus, more than one gene. Furthermore, the mode of synthesis of miRNA and siRNA is different. miRNA is synthesized endogenously from cell DNA in a two-step process ending in the nucleus and cytoplasm, while siRNA is added exogenously to the cell. Since miRNA can regulate the expression of multiple mRNAs and miRNAs are involved in cancer physiology, they have been recently explored as both therapeutics and targets for cancer treatment. Studies involving miRNA therapeutics are increasing, as there are currently approximately 160 active clinical trials (www.clinicaltrial.gov).

Various nanoparticle formulations have been developed for in vitro and in vivo delivery of miRNAs and anti-miRNAs (antimirs) for therapeutic purposes. However, the challenge of miRNA delivery is the lack of a specific, nontoxic, and efficient method of delivery. Since miRNA can also act as an oncoMiR, an antimir is often used to suppress the expression of such oncogenic miRNA. miR-21 is an oncoMir in many cancers, including triple-negative breast cancer (TNBC). In a recent study by Shu et al., an RNA nanoparticle conjugated with epidermal growth factor receptor (EGFR) aptamer was used to target delivery of antimirs for miR-21 in TNBC, in vitro and in vivo (Shu et al., 2015). Ultrasound-induced microbubble cavitation has been widely recognized as a safer way of delivering therapeutics. Wang et al. used ultrasound cavitation to deliver miR-122 loaded with PLGA–PEG nanoparticles into human colon cancer xenografts in mice. The findings showed that this novel approach can be used for noninvasive delivery of therapeutic miRNA in cancer treatment (Wang et al., 2015).

Multiple miRNAs often work together as clusters that are produced from a single pri-miRNA. In a study by Subramanian et al., an aptamer for the miT17~92 cluster was used in retinoblastoma cell lines and resulted in downregulation of miRNA of the cluster (Subramanian, Kanwar, Kanwar, & Krishnakumar, 2015). In addition, various conventional metallic nanoparticles, such as galadonium, gold, and iron, have been found to functionalize with miRNA and have been extensively explored (Hsu et al., 2013; Yoo et al., 2014). Recently, the presence of miRNAs was reported in the lumen of cellular vesicle, such as exosomes, which are already known for carrying and delivering cellular cargo across the cellular microenvironment. In recent years, the miRNA present in exosomes has been exploited as a therapeutic drug delivery vehicle (Srivastava, Babu, et al., 2016; Srivastava et al., 2015). In addition, loading exosomes with miRNA of therapeutic importance has been reported. Researchers have also adopted the use of biomimetics, which are mimics of natural vectors and have similar properties. Although the concept of miRNA as gene therapy molecule is relatively new, this multitargeted approach is expected to become a popular tool for cancer treatment and a novel approach for efficient drug delivery.

## 4. NANOPARTICLE-BASED CODELIVERY OF DRUGS AND GENES

Individual cancer therapeutic agents and approaches may not sufficient to cure cancers. Using high doses of drug to increase the therapeutic effect may lead to drug resistance and undesirable side effects (Liu, 2009). Therefore, combined therapy may be more effective than the corresponding individual treatments alone. The rationale to use combination therapies is that the therapeutics work by different mechanisms of action, which leads to reduced resistance in cancer cells (Herman et al., 1988). In combination therapy, each therapeutic agent can be used at the optimal dose that is tolerable to healthy tissues to reduce toxicity. Based on the type and stage of cancer, the appropriate combination therapy will be used. Combination therapies are more suitable for advanced cancers, such as nonsmall cell lung cancer, esophageal cancer, or bladder cancer.

Since cancer is a complex disease involving multiple targets of therapy, the codelivery of a chemotherapeutics with genes has been explored. Chemotherapy is one of most used approaches in cancer treatment; however, it has some drawbacks, such as drug resistance and toxicity to healthy tissues (Liu, 2009). RNAi has been incorporated into nucleic acid medicines for cancer treatment. The codelivery of nucleic acids with chemotherapy drugs can have additive or synergistic effects that maximize therapeutic efficacy. Codelivery overcomes many obstacles of individual treatments, such as drug resistance and toxicity to normal tissues (Wang, Zhao, et al., 2010). However, a safe and efficient delivery system is needed for the codelivery of nucleic acids and chemotherapeutic drug molecules. Nanoparticle-based delivery systems are excellent carriers and have been extensively studied in this context. This section discusses the codelivery of RNAi molecules with chemotherapy drugs via different types of nanoparticles (Khan, Ong, Wiradharma, Attia, & Yang, 2012).

Zhu et al. developed binary polymer low-density lipoprotein-*N*-succinyl chitosan–cystamine–urocanic acid (LDL-NSC-SS-UA) micelles with dual pH/redox sensitivity and targeting for the codelivery of breast cancer resistance protein, siRNA, and paclitaxel (PTX). These siRNA–PTX-loaded micelles exhibit stability under physiological conditions. In the tumor microenvironment (pH/redox), the micelles showed fast release of gene and drug molecules. These micelles showed enhanced antitumor activity and downregulated the protein and mRNA expression levels of breast cancer resistance protein in MCF-7/Taxol cells. The in vivo study results revealed that the siRNA–PTX-loaded micelles showed prolonged circulation time with a remarkable tumor-targeting effect, and effectively inhibited tumor growth (Zhu et al., 2017). Another group reported a redox-sensitive micellar system that was synthesized with hyaluronic acid-based amphiphilic conjugate HA-ss-(OA-g-bPEI), HSOP, and was loaded with paclitaxel (PTX) and AURKA-specific siRNA (si-AURKA) with tumor-targeting molecules. In vitro and in vivo, the HSOP micelles simultaneously delivered the PTX and siRNA, producing synergistic effects between the drugs that led to greater antitumor efficacy than that observed with single drug-loaded micelles and nonsensitive coloaded micelles (Yin, Wang, Yin, Zhou, & Huo, 2015).

IL17RB siRNA and DOX were codelivered using a chitosan-based nanoparticle and showed enhanced anticancer efficacy. Similarly, VEGF and Bcl-2 dual-targeted siRNA was used to study prostate cancer, and the results revealed enhanced treatment efficacy when the dual siRNA was administered (Lee et al., 2015). In theory, such an approach can be adopted in which the drug resistance gene is silenced, followed by the delivery of drug. Similarly, a lipid–polymer nanoparticle system was used to deliver gefitinib and HIF1α gene treatment for pancreatic cancer (Zhao et al., 2015). The use of chitosan increases the functionality of codelivery nanoparticle systems. For instance, Babu et al. (2014) prepared a nanoparticle system based on chitosan and PLA for codelivery of P62 siRNA, proteasome β5 plasmid, and cisplatin. The nanoparticle was designed such that the PLA nanoparticle encapsulated the drug cisplatin and was coated with chitosan, onto which sip62 and pβ5 were adsorbed by maintaining a siRNA:pDNA:CS ratio of 1:0.2:53 (w/w). This codelivery strategy using a chitosan–PLA nanoparticle system resulted in P62 knockdown, β5 overexpression, and enhanced cisplatin sensitivity in drug-resistant ovarian cancer cells. Fig. 2 shows the effect of P62 and β5 protein expression levels in cisplatin-resistant 2008/C13 cells that received various treatments using the chitosan–PLA codelivery system (Babu et al., 2014).

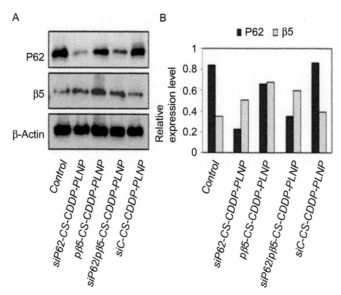

**Fig. 2** (A) The effect of codelivery of siP62 and/or pβ5 using cisplatin-encapsulated, chitosan-coated polylactic acid nanoparticles (CS-CDDP-PLNP) on P62 and β5 protein expression levels in cisplatin-resistant 2008/C13 cells. (B) The corresponding diagram shows relative expression levels of the two proteins normalized to beta actin. *Figure reproduced from Babu, A., Wang, Q., Muralidharan, R., Shanker, M., Munshi, A., & Ramesh, R. (2014). Chitosan coated polylactic acid nanoparticle-mediated combinatorial delivery of cisplatin and siRNA/plasmid DNA chemosensitizes cisplatin-resistant human ovarian cancer cells. Molecular Pharmaceutics, 11(8), 2720–2733. doi:10.1021/mp500259e. Copyright © 2014, American Chemical Society.*

Different nanoparticles with drug combinations have gained popularity recently. In a solid tumor model, shRNA against thymidylate synthase (TS shRNA) complexed with cationic liposome (TS shRNA-lipoplex) was codelivered with oxaliplatin and showed enhanced effects and accumulation of the TS shRNA lipoplex. In breast cancer, a combination treatment using doxorubicin functionalized to graphene oxide (GO-PAMAM) with shRNA against MMPS showed an enhanced therapeutic response (Gu et al., 2017). Zhang et al. used a combined approach to deliver Akt-shRNA with drug demethylcantharate. The results in three cancer cell lines confirmed the enhanced therapeutic response of this combination molecule (Zhang et al., 2016).

Codelivery of miRNA (or antimiRs) and drugs is expected to produce an enhanced effect, as miRNA can affect the expression of several genes involved in the signaling pathways, which can enhance the sensitivity of cell for a given drug. Similarly, tumor suppressor miR can be delivered to curtail the effect of metastatic relapse and rapid cell proliferation, shielding the cells from the effects of the drug. A combined formulation of DOX and miR 34a in breast cancer produced enhanced antitumor effects in breast cancer in vivo and in vitro. The therapeutic effect of DOX is further complemented with miR-34a, which inhibits cancer cell migration by targeting the Notch-1 signaling pathway. A hyaluronic acid–chitosan nanoparticle was formulated for delivery of this combination (Deng et al., 2014). Another approach to achieve synergistic therapeutic response is to use multiple miRNAs involved in a critical pathway (Pencheva et al., 2012).

## 5. STIMULI-RESPONSIVE DRUG DELIVERY

Recent progress in materials science and drug delivery allows spatial-, temporal-, and dosage-controlled mechanisms to be introduced in nanoparticle drug delivery systems. Such a modification to nanoparticles can be achieved by using stimuli-responsive materials in their synthesis. The stimuli-response delivery systems address the issues of controlled dose release of drug in response to various stimuli signals specifically produced in the tumor microenvironment, as represented in Fig. 3. Therefore, there will in theory be almost zero drug release until stimuli are applied. Major stimuli signals in the tumor microenvironment include pH, redox state and concentration, and types and concentrations of proteins and enzymes

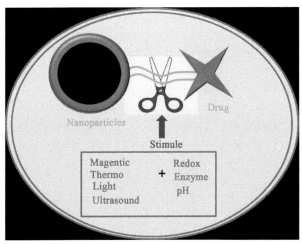

**Fig. 3** Schematic representation illustrating stimuli-responsive, tumor-targeted drug delivery under extracellular and intracellular gradients in nanoparticle–drug conjugates.

(Lopes, Santos, Barata, Oliveira, & Lopes, 2013). The response of nanoparticle drug delivery systems to these internal stimuli allows tumor-specific release of the payload. External stimuli can also be applied for additional specificity in tumor-targeted drug delivery (Puoci, Iemma, & Picci, 2008). In the following subsections, we focus on recent developments in the design of nanoparticle-based stimuli-responsive systems that are able to control drug release in response to external (temperature, magnetic field, ultrasound, light, or electric pulses) or internal (pH, enzyme concentration, or redox gradients) stimuli.

## 5.1 Thermostimuli

Thermoresponsive drug delivery systems are more stable at body temperature, 37°C. However, with a slight increase in temperature to 40–42°C, the carrier exhibits stimuli responsiveness and released the payload (Bikram & West, 2008). Usually, liposomes and polymers respond to changes in the temperature of the external environment, which then changes their conformation, solubility, and hydrophilic/hydrophobic balance (Kneidl, Peller, Winter, Lindner, & Hossann, 2014). The most commonly used thermosensitive polymer is poly(N-isopropyl acrylamide), commonly called PNIPAM. PNIPAM exhibits a lower critical solution temperature (LCST) parameter (Gong et al., 2013).

Tagami et al. reported the development of thermoresponsive liposomes (HaT liposome) for image-guided drug delivery in which they codelivered Gd-DTPA for T1 MRI and chemotherapy drug doxorubicin (DOX). They observed that 100% of Dox was released at 40–42°C in 3 min with a correlation of MR T1 values that indicated a 60% reduction in MR T1 relaxation values. In the in vivo study, researchers observed that DOX showed more antitumor efficacy with enhancement of T1 signal in a heated tumor environment (Tagami et al., 2011). Another group utilized the dissociative properties of thermosensitive liposomes, which combine with unfolding advantages of temperature-sensitive peptide to improve DOX accumulation in tumors with leucine zipper peptide–liposome hybrid nanocarriers (Al-Ahmady et al., 2012). Another hydrophobic chemotherapy drug, paclitaxel (PTX), also showed enhanced antitumor activity in vitro and in vivo when delivered using thermosensitive liposomes. Wang and coworkers reported more PTX release at 42°C than at 37°C. An in vivo study of a xenograft lung tumor model showed enhanced drug release in tumor tissues when hyperthermia occurred, resulting in increased tumor growth suppression, compared with nontemperature-sensitive liposome and free drug treatment (Wang, Zhang, et al., 2016).

Polymers are another important class of thermosensitive materials. Among these, poly(N-isopropylacrylamide-*co*-acrylamide), commonly called P(NIPA-*co*-AAm), is extensively utilized. Zhang et al. studied P(NIPA-*co*-AAm) as a hydrogel in delivering multiple hydrophobic drugs, 5-fluorouracil (5-FU), fluorescein, docetaxel (DTX), and near-infrared dye-12 (NIRD-12), and performed in vitro release studies at different temperatures. DTX-loaded thermoresponsive hydrogels enhanced the tumor inhibition to 78.15% with hyperthermia in Kunming mice bearing S180 sarcoma compared with the same mice without hyperthermia (48.78%), and reduced toxicity to normal tissues (Zhang, Qian, & Gu, 2009). Another polymer, poly($\gamma$-2-(2-(2-methoxyethoxy)-ethoxy) ethoxy-$\varepsilon$-caprolactone)-*b*-poly($\gamma$-octyloxy-$\varepsilon$-caprolactone), PMEEECL-*b*-POCTCL, exhibits LCST at 38°C and demonstrated thermoresponsive properties. Dox-loaded thermoresponsive micelles of PMEEECL-*b*-POCTCL showed increased cytotoxicity at the above LCST in the MCF7 breast cancer cell line (Cheng et al., 2012).

## 5.2 Magnetic Stimuli

Magnetic nanoparticles convert an applied magnetic field into heat. Because of the transformation of magnetic energy into heat by the dynamic response of a dipole with their magnetic moments, that heat acts as a stimulus in delivering the drugs (Giri, Trewyn, Stellmaker, & Lin, 2005). Magnetic-responsive drug delivery has great potential for the development of safe and dual therapy agents (Kumar & Mohammad, 2011). The external magnetic field plays an important role in the controlled release of drugs from magnetic drug delivery carriers.

Magnetic nanoparticles must be modified with other materials to load drugs for stimuli-responsive delivery. Zhang et al. developed a fluorescent-magnetic biotargeting system with PEI-modified fluorescent $Fe_3O_4$@$mSiO_2$ yolk-shell nanobioprobes as a model for simultaneous fluorescence imaging and magnetically guided drug delivery in liver cancer cells (Zhang et al., 2012). Another group developed poly[aniline-*co*-N-(1-one-butyric acid) aniline] (SPAnH) coated on $Fe_3O_4$ cores, which enhanced the therapeutic capacity and thermal stability of 1,3-bis (2-chloroethyl)-1-nitrosourea (BCNU), a compound used to treat brain tumors. Both in vitro and in vivo studies showed that, after applying an external magnetic field, nanoparticles bound-BCNU-3 accumulated at targeted tissues and increased the therapeutic efficacy, while reducing toxicity (Hua et al., 2011).

In another approach, Harries and coworkers demonstrated stimuli-responsive controlled delivery of vancomycin antibiotic with $Fe_3O_4$ nanoparticles. Both the antibiotic and $Fe_3O_4$ nanoparticles were loaded onto chitosan microbeads with polyethylene glycol dimethacrylate as the crosslinker. Researchers studied the elution properties with successive magnetic stimulations at multiple field strengths and frequencies. They observed that 30 min of magnetic stimulation induced a temporary increase in the daily elution rate of up to 45%. This increase was influenced by field strength, field frequency, and crosslinker length. After 3 days, 70% of the beads were degraded, but continued to elute antibiotic for up to 8 days. No cytotoxic effects were observed in vitro, compared with controls (Harris et al., 2017). Recently, another group reported using the $MnFe_2O_4$ core and NIR/pH-coupling sensitive mesoporous silica shell nanocarriers for enhanced antitumor efficacy and $T_1/T_2$-weighted dual-mode MRI applications in vivo. This nanocomposite showed the photothermal/chemo dual-modal synergistic therapies triggered by NIR/pH. They observed that, under 808 nm irradiation, $MnFe_2O_4$ can transform light into heat, which ablates tumor cells directly and also promotes the release of DOX from the mesoporous layer (Chen, Deng, et al., 2017). Deng and colleagues reported the development of a dual, external triggered, and magnetic stimuli-responsive delivery system with hybrid microcapsule (h-MC) by an electrostatic layer-by-layer assembly of polysaccharides (sodium alginate, chitosan, and hyaluronic acid), iron oxide, and graphene oxide (GO), in which the drug is loaded through pH control. The alternative irradiation of magnetic field and NIR laser applied on triggerable $Fe_3O_4$@GO component for dual high-energy and high-penetration hyperthermia therapy (Deng et al., 2016).

## 5.3 Light Stimuli

Drug release from light-responsive drug delivery systems can be controlled temporally and spatially using specific light irradiation of different wavelengths. Delivery systems have been developed that respond to light of ultraviolet (UV), visible, or NIR wavelengths (Alvarez-Lorenzo, Bromberg, & Concheiro, 2009). Light-induced delivery systems can be achieved through different mechanisms, such as light-induced isomerization, bond cleavage, and disaggregation of materials (Fomina, Sankaranarayanan, & Almutairi, 2012). Drugs can be conjugated with nanoparticles through photocleavable ligands. Upon light irradiation, ligands cleave or activate and release the drugs.

Zhang et al. reported a titanium dioxide-based visible light-sensitive nanoplatform (HA–TiO$_2$–IONPs/ART) for codelivery of artemisinin as a type of Fe$^{2+}$-dependent drug and Fe$^{2+}$ cotransport system. This system showed improved antitumor effects. The researchers observed that the HA–TiO$_2$–IONPs/ART system synchronized the codelivery of Fe$^{2+}$ and artemisinin tumor-responsive release, and generated ROS under visual light irradiation. Both in vitro and in vivo results indicated that the HA–TiO$_2$–IONPs/ART system exhibits enhanced antitumor activity when combined with laser irradiation (Zhang, Zhang, et al., 2017). This switchable dual light- and temperature-responsive drug carrier system was reported to use a gold nanoparticles (Au NPs)-grafted poly(dimethylacrylamide-co-acrylamide)/poly(acrylic) acid [P(DMA-co-AAm)/PAAc] hydrogel. Then, the researchers studied the swelling, thermal sensitivity, and thermal and optical switching properties of the prepared hydrogels in acidic (pH 1.2) and neutral (pH 7.4) buffered solutions to simulate stomach and intestinal conditions. During ofloxacin antibiotic release studies, it was observed that the "on" state occurred at higher temperatures and the "off" state occurred at lower temperatures (Amoli-Diva, Sadighi-Bonabi, & Pourghazi, 2017) under thermal and optical switching conditions.

Li et al. demonstrated a CDDP drug delivery system with NIR light stimuli-responsive drug release properties based on the coordination chemistry of CDDP and tellurium-containing block polymer (PEG–PUTe–PEG). This nanocarrier was loaded with indocyanine green (ICG) as an NIR-sensitive dye. Under NIR laser irradiation, ROS was generated from ICG. The ROS oxidizes the tellurium. The oxidation of tellurium weakened the coordination bond between CDDP and tellurium and led to the rapid release of CDDP. In addition, the researchers observed that the encapsulated ICG had a synergistic antitumor effect through photothermal effects after mild laser irradiation (Li, Li, Cao, Wang, & Xu, 2017). Another group reported the use of arylboronic ester and cholesterol-modified hyaluronic acid (PPE–Chol$_1$–HA), denoted as the PCH–DI nanosystem, for light-triggered DOX release, along with ICG PTT. The ICG produced ROS that cleaved arylboronic ester to realize controllable drug release. The results were confirmed by NIR laser irradiation. The in vitro results indicated that DOX in the PCH–DI/laser group showed the most efficient nuclear binding toward HCT-116 colon cells; the results of the in vivo studies indicated that this system also enhanced the cytotoxicity and tumor suppression effect of PCH–DI on a nude mouse model of HCT-116 tumor xenografts (Chen, Deng, et al., 2017).

## 5.4 Ultrasound Stimuli

Ultrasound-sensitive drug delivery systems can improve targeted delivery of drugs into specified tissues, while reducing the systemic dose and toxicity (Paliwal & Mitragotri, 2006). Sonoporation is an important mechanism for stimuli-enhanced delivery of therapeutics in this system through vasculature openings, induced by ultrasound-triggered oscillations and destruction of microbubbles (Ferrara, 2008).

Ultrasound interaction with nanoparticles induces enhanced drug delivery and can also be used in image-guided delivery. Chen et al. developed an ultrasound imaging contrast and ultrasound-triggered DOX using polymer microcapsules composed of hydrogen-bonded multilayers of tannic acid and poly(N-vinylpyrrolidone). These capsules work as diagnostic agents upon low-power ($\sim 100\,\text{mW/cm}^2$) ultrasound irradiation and act as therapeutics upon high-power ($>10\,\text{W/cm}^2$) ultrasound irradiation. The researchers observed that DOX release was gradual at lower power and swift at higher power. In vitro results indicated that 50% of DOX release induced 97% cytotoxicity in MCF-7 breast cancer cells, whereas no cytotoxicity was found in the absence of ultrasound irradiation (Chen, Ratnayaka, et al., 2017).

Another group developed multifunctional smart curcumin-loaded chitosan/perfluorohexane nanodroplets for ultrasound imaging mediated, on-demand drug delivery systems. The researchers observed that 63.5% of curcumin was released from the nanoformulation upon sonication at the frequency of 1 MHz, $2\,\text{W/cm}^2$ for 4 min. Cell growth inhibition was significantly increased in curcumin-loaded nanodroplets of 4T1 human breast cancer cells upon ultrasound exposure (Baghbani, Chegeni, Moztarzadeh, Hadian-Ghazvini, & Raz, 2017). Another report from same group demonstrated that codelivery of doxorubicin and curcumin using multifunctional smart alginate/perfluorohexane nanodroplets (Dox-Cur-NDs) had a synergistic therapeutic effect in adriamycin-resistant A2780 ovarian cancer cells for Dox-Cur-NDs with ultrasound irradiation compared to Dox-NDs; in vivo results also demonstrated efficient tumor suppression (Baghbani & Moztarzadeh, 2017).

## 5.5 Redox Stimuli

Biologically reduced substances vitamin C (ascorbic acid), vitamin E, and glutathione (GSH) are widely distributed in all body tissues (Sun, Meng, Cheng, Deng, & Zhong, 2014). Reduced state GSH plays an important role in human metabolism. In intratumoral conditions, GSH expression is almost a 1000-fold greater than in normal tissues (Wen et al., 2011). Hence, this property has motivated researchers to fabricate redox-sensitive drug delivery systems using disulfide crosslinkers.

The redox-controlled release of drugs can revert MDR in tumors, with fewer side effects. Qiao et al. created a D-α-Tocopherol polyethylene 1000 succinate (TPGS)-based drug delivery system that contains both TPGS and mitoxantrone (MTO) via a disulfide bond and assembles into micelles (TSMm). The disulfide bonds are more sensitive at high concentrations of intracellular glutathione (GSH), which lead to rapid release of MTO. These redox-sensitive TSMm showed significantly increased therapeutic effect in treatment-resistant MDA–MB-231/MDR breast tumor cells, compared with either free MTO or disulfide-free prodrug micelle (TCMm). Furthermore, TSMm has demonstrated significantly stronger antitumor activity in xenograft nude mice, without causing toxicity (Qiao et al., 2017).

PTX is highly effective in many cancer types, but its highly hydrophobic nature limits the usage in current formulations. Redox-sensitive PTX delivery overcomes this obstacle. Li and coworkers developed a nanoformulation with α-amylase- and redox-responsive nanoparticles based on hydroxyethyl starch (HES) for the redox-sensitive delivery of PTX. These HES-SS-PTX nanoparticles showed a prolonged half-life compared with that of the commercial PTX formulation (Taxol), resulting in more tumor accumulation. In reductive conditions, HES-SS-PTX NPs showed better therapeutic efficiency than did Taxol in in vitro and in vivo models in 4T1 tumor-bearing mice (Li, Hu, et al., 2017). In another approach, a mixed micelle system was developed with redox-sensitive mPEG-SS-PTX and mPEG-SS-DOX conjugates to reduce the side effects. The in vitro release profile and in vitro anticancer activity of mixed micelles showed redox-controlled release and significant cytotoxicity in A549 and B16 cells. The in vivo results demonstrated that the mixed micelles had fewer side effects than free PTX/DOX in mice (Zhao et al., 2017).

Pluronic F127 (F127), an amphiphilic triblock copolymer, improved the drug delivery efficiency by introducing a redox-sensitive disulfide linker into micelles. Liu et al. synthesized α-tocopherol (TOC) conjugated with F127 polymer (F127-SS-TOC) through a redox-sensitive disulfide bond between F127 and TOC. These F127-SS-TOC micelles showed high hemocompatibility and low cytotoxicity in Bel 7402 and L02 cells (Liu, Fu, et al., 2017).

## 5.6 Enzyme Stimuli

Biochemists have explored the role of biomolecules in biochemical reactions and their distribution within the body. It is well established that enzymes are involved in most of the biocatalytic reactions in the body (de la Rica, Aili, & Stevens, 2012). Enzymes exhibit properties that catalyze the biodegradation of biomacromolecules; however, some enzymes are overexpressed in cancer cells (Hu, Katti, & Gu, 2014). Enzyme-degradable linker molecules can be used to make enzyme-responsive drug delivery systems. For example, Zhang et al. developed an enzyme-responsive PEGylated lysine peptide dendrimer–gemcitabine conjugate (dendrimer–GEM)-based nanoparticle through a click reaction. Glycyl phenylalanyl leucyl glycine tetra-peptide (GFLG) was used as an enzyme-cleavable linker. The nanoparticle showed significantly rapid GEM release in the secreted Cathepsin B environment, compared with conditions without Cathepsin B. This dendrimer–GEM nanoparticle displayed enhanced antitumor efficacy in a 4T1 murine breast cancer model, with minimal toxicity; twofold higher tumor growth inhibition was induced by the dendrimer–GEM nanoparticle than by free GEM (Zhang, Pan, et al., 2017).

In addition, there was a recent report of an enzyme-responsive drug delivery system based on enzymatic biodegradable amphiphilic poly(ester-urethane)s from L-tyrosine amino acid resources. The self-assembled nanoparticles were used to deliver anticancer drugs DOX and camptothecin (CPT). These nanoparticles showed more intracellular drug release than extracellular drug release, due to enzymatic biodegradation. The cell uptake and in vitro cytotoxicity studies demonstrated that the L-tyrosine nanoparticles showed more cell internalization and cell killing efficiency in HeLa cervical cancer cells, with reduced toxicity to normal cells (Aluri & Jayakannan, 2017).

Kumar and colleagues used guar gum, a natural carbohydrate polymer, as a capping layer to encapsulate chemotherapy drug 5-flurouracil (5FU) within the channels of MSNs for oral delivery of a colon-specific drug. They observed that the specific release of 5FU from guar gum-capped MSN (GG-MSN) occurs via enzymatic biodegradation of guar gum by enzymes in the colonic microenvironment. However, there was no release in the gastrointestinal tract. The released drug showed anticancer activity in colon cancer cell lines (Kumar et al., 2017).

## 5.7 pH Stimuli

Owing to enhanced glycolysis and accumulation of lactic acid, the tumor microenvironment (endo/lysosomes) has a lower pH gradient (4–6.5) than that (7.4) of the corresponding normal tissues (Estrella et al., 2013). pH-sensitive drug delivery has been shown to overcome the MDR and reduce toxicity to normal cells. Furthermore, pH-responsive drug delivery systems are mostly created with pH-sensitive linkers (hydrazone and ester linkages) that are conjugated at one end with chemotherapy drug and at the other end with nanocarriers (Prabaharan, Grailer, Pilla, Steeber, & Gong, 2009).

Ir (III) drugs have potent anticancer activity. Cai and colleagues examined the pH- and redox-sensitive delivery of these Ir (III) drugs. The researchers developed CD44-targetable conjugates of HA-cystamin-pyrenyl (HA-ss-Py) containing disulfide bonds, and HA-pyrenyl (HA-Py), which forms micelle particles (Cai et al., 2017). Almeida et al. developed a method of pH- and thermoresponsive curcumin (CUR) delivery of magnetic microgels using pectin maleate, N-isopropyl acrylamide, and $Fe_3O_4$ nanoparticles. CUR was loaded into the microgels, and slow and sustainable CUR release was achieved under the influence of an external magnetic field (Almeida et al., 2017). Qiu et al. reported novel tumor-targeted hyaluronic acid-2-(octadecyloxy)-1,3-dioxan-5-amine (HOD) conjugates. The HOD micelles were sensitive to degradation in the acidic tumor microenvironment. Then, DOX was loaded on HOD micelles (DOX/HOD), which showed enhanced release at low pH. The DOX/HOD micelles exhibited enhanced cell killing efficiency, compared with pH-insensitive micelles and the free DOX control (Qiu et al., 2017).

Our group also extensively studied pH-sensitive DOX release using nanosomes as carriers. The nanosomes were made up of gold nanoparticles and exosomes. We conjugated DOX to gold nanoparticles through a hydrazone-based, pH-sensitive drug delivery linker. Then, the nanoparticles were loaded on exosomes that were isolated from normal MRC9 lung fibroblast cells. Nanosome formation was confirmed by transmission electron microscopy (TEM) and zeta potential, as represented in Fig. 4. The highest DOX release was observed in conditions that mimicked the tumor microenvironment. Increased cell uptake and cytotoxicity were observed in lung

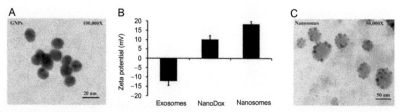

Fig. 4 (A) TEM image of gold nanoparticles, (B) surface zeta potential values of exosomes that were isolated from MRC9 cells, NanoDox (GNP–DOX), nanosomes (exosomes+GNP–DOX), and (C) TEM image of nanosomes. *Figure reproduced from Srivastava, A., Amreddy, N., Babu, A., Panneerselvam, J., Mehta, M., Muralidharan, R., ..., Ramesh, R. (2016). Nanosomes carrying doxorubicin exhibit potent anticancer activity against human lung cancer cells.* Scientific Reports, 6, 38541. doi:10.1038/srep38541. *Copyright © 2016, Rights Managed by Nature Publishing Group.*

cancer cell lines, while toxicity to normal fibroblast cells and DOX-sensitive human coronary artery smooth muscle cells were reduced (Srivastava, Amreddy, et al., 2016).

The above examples of stimuli-responsive nanoparticle systems demonstrate the growing importance for these systems in drug delivery applications. However, most are still in the early stages of development. Thorough optimization of the synthesis procedures is needed before these systems can transition into clinical settings. Many of the above-discussed systems failed to surpass in vitro levels, but a few have entered in vivo trials. Simplified systems with strong stimuli-responsive characteristics may drastically influence the chances for clinical applications. Despite these hurdles, stimuli-responsive drug delivery systems have the potential to replace current drug delivery approaches soon.

## 6. NANOPARTICLE-BASED RECEPTOR-TARGETED DELIVERY

Translation of cancer therapeutics and gene therapy into the clinic depends on the successful development of targeted delivery systems. Conventional chemotherapeutic drugs cause several side effects, including nonspecific killing of both cancerous and noncancerous cells (Morrow, 1985). These findings led to advancements in the development of molecularly targeted nanocarriers for selective uptake of the therapeutic molecules in the tumor site (Brannon-Peppas & Blanchette, 2004). These targeted nanocarriers protect the therapeutics from serum proteins, increase their systemic circulation, and enhance their efficacy, while minimizing specific toxicity. An efficient targeted delivery system should remain in the systemic circulation until it reaches the tumor site, while retaining its characteristics and releasing the therapeutics at the tumor site. Once the nanocarriers are injected into the systemic circulation, they will reach the tumor site through active or passive targeting mechanisms. In passive targeting, the nanocarriers reach the tumor site as a result of enhanced permeability and retention phenomena, known as the EPR effect; which depends on the size of the nanocarriers, and leaky vasculature and impaired lymphatics. Passive targeting is influenced by several independent properties that affect the delivery system, such as size, shape, surface charge, surface chemistry, composition, and other physicochemical properties (Yu et al., 2016). In active targeting, the nanocarriers reach the tumor site based on the affinity of the ligand toward the antigens or receptors that are highly expressed on the tumors.

The approach of targeting biologically active pathways to produce synergistic effects without causing any toxic effects leads to successful translation (Wang, Yu, Han, Sha, & Fang, 2007). Therefore, identifying several targeted molecules and designing nanocarriers became desirable in the field of targeted delivery systems. Multifunctionality can be improved if combined with receptor-mediated targeted delivery by conjugating targeting ligands on the surface of nanocarriers, as illustrated in Fig. 5. Here, we discuss the recent examples of EGFR-, transferrin-, luteinizing hormone-releasing hormone (LHRH) receptor-, folate receptor- (FR), and Arg-Gly-Asp (RGD) receptor-targeted therapeutic delivery with nanocarriers, and their anticancer activity.

## 6.1 EGFR Receptor

EGFR is a transmembrane tyrosine kinase receptor that is overexpressed in many types of cancer, including lung, breast, ovarian, and pancreatic cancers (Inicholson, WGee, & Harper, 2001). EGFR plays a major role in regulating cell proliferation and survival through three different pathways: RAS–RAF mitogen-activated protein, PI3K/AKT, and JAK/STAT (Okayama et al., 2012).

A polymer-based unimolecular NP formulation with pH/redox dual stimuli release of siRNA for EGFR-targeted delivery was recently reported (Chen, Wang, Xie, & Gong, 2017). The researchers utilized GE11 peptide for EGFR targeting, and it overcame the limitations associated with endosomal and lysosomal escape of siRNA. In this study, siRNA was

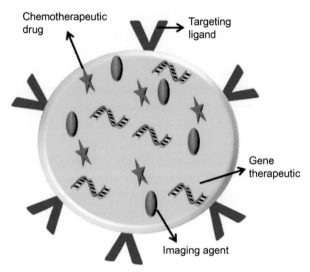

**Fig. 5** Schematic illustration of multifunctional nanoparticles for tumor-targeted delivery of therapeutics and imaging agents using targeted ligands functionalized on the surface of nanocarriers.

complexed at the multiarm cationic core polymer through electrostatic interaction. An imidazole group in the cationic polymer absorbed the protons from acidic endosomes, leading to osmotic swelling and facilitating the endosomal escape of the siRNA complex. These pH/redox dual-sensitive nanoparticles were functionalized using EGFR-targeting GE11 peptide to enhance the active targeting and efficient gene silencing in GFP-MDA-MB-468 TNBC cells, without any significant cytotoxicity (Chen, Wang, et al., 2017). Researchers developed superparamagnetic iron oxide (SPIO) nanoparticles for MRI sensitivity and energy deposition efficiency using a clinical MRgFUS system. The SPIO surface was decorated with EGFR monoclonal antibody (mAb) to target EGFR overexpressed in lung cancer cells. They demonstrated that the anti-EGFR, mAb-targeted SPIO nanoparticles significantly improved the targeting efficiency in H460 lung tumor cells, and this MRI contrast imaging facilitated the monitoring of therapeutic efficiency in a nude rat model (Wang, Qiao, et al., 2017).

Later, the same group reported Cetuximab-conjugated micelles loaded with IR-780 (radiolabeled with Indium-111) for targeted imaging and PTT. These micelles specifically bind to EGFR at the tumor site and showed enhanced therapeutic effect. Micro-SPECT/CT imaging and biodistribution studies showed the accumulation of Cetuximab micelles at the tumor. The radiolabeled radioisotopes provided real-time monitoring of the micelles in the circulation. This approach can be used for optical and nuclear imaging, and to deliver therapeutics simultaneously (Shih et al., 2017). Studies from our lab demonstrated the conjugation of anti-EGF antibody into plasmonic resonance magnetic nanoparticles, which were selectively taken up by EGFR-overexpressing lung tumor cells. This EGFR-targeted nanoparticle showed enhanced antitumor activity by inducing apoptosis and autophagy in NSCLC cells (Yokoyama et al., 2011).

## 6.2 Transferrin Receptor

Transferrin receptor (TfR1), also known as CD71, is ubiquitously expressed at low levels in most human tissues. It is expressed on the outer cell membrane and transports iron into cells through a clathrin/dynamin-dependent endocytotic mechanism (Daniels et al., 2012). Higher expression of TfR1 in malignant cells is correlated with tumor progression, and make this receptor an attractive target for delivery of nanotherapeutics. Direct conjugation of chemotherapeutic drugs to transferrin showed enhanced and selective delivery in malignant cells. Covalent linkage of transferrin to doxorubicin showed cytotoxicity in multiple cancer cell lines (Szwed, Kania, & Jozwiak, 2014). Surface functionalization of nanoparticles using transferrin peptide, antibody, or protein provided promising results. Conjugation of transferrin as a targeting moiety on the surface of the nanoparticle showed enhanced delivery due to the availability of the receptors on malignant cells. Sahoo et al. developed PLGA nanoparticles loaded with paclitaxel and conjugated to transferrin, and tested the nanoparticles in PC3 human prostate cancer cells. Systemic administration of these nanoparticles resulted in complete regression of the tumor and increased survival in nude mice bearing PC3 cells (Sahoo, Ma, & Labhasetwar, 2004). A pulmonary siRNA delivery system using Tf-PEI for asthma therapy showed selective delivery of siRNA to activate T cells in the lung. In vitro studies demonstrated the enhanced uptake of these polyplexes in human primary activated T cells and JURKAT cells. Biodistribution studies in murine asthma model showed selective delivery of siRNA to activated T cells (Xie, Kim, et al., 2016).

Researchers have shown that the cationic polymer branched polyethyleneamine (bPEI) coupled with TfR targeting peptides (HAIYPRH peptide) efficiently delivered siRNA into TfR-overexpressing lung cancer cells. The targeted delivery resulted in GAPDH gene knockdown that was more efficient than that observed with bPEI alone (Xie, Killinger, Moszczynska, & Merkel, 2016).

Studies from our lab and others have shown that the RNA-binding protein HuR regulates several target mRNAs that are involved in cancer progression; thus, HuR has emerged as a promising target for cancer treatment. The target genes of HuR are key regulators of the cell cycle, angiogenesis, and cell migration and invasion. Hence, we investigated whether siRNA-mediated knockdown of HuR would induce cell death, leading to better survival. To test this approach, we developed an efficient lipid nanoparticle delivery system targeting transferrin receptors (TfNP) that are overexpressed on lung tumor cells. Our in vitro studies demonstrated the selective uptake of TfNP, which led to downregulation of HuR and HuR-regulated oncogenes. Biodistribution studies revealed the selective accumulation of TfNPs at the tumor site. In vivo efficacy studies showed significant reductions in xenograft tumor volumes and in the number of lung met tumor nodules upon intravenous administration of TfNP into nude mice (Muralidharan et al., 2017).

Liposomal nanocarriers have gained popularity for applications as drug carriers. We attempted to develop a liposomal nanocarrier to enhance the delivery of tyrosine kinase inhibitor gefitinib to a lung tumor model. We successfully loaded the poorly soluble drug gefitinib into the lipid bilayer of a liposomal structure. These gefitinib-loaded liposomes delivered a combination of gefitinib and HuR siRNA. This system demonstrated the combined inhibitory activity of gefitinib and HuR siRNA in the tested lung tumor cells, which were analyzed by Western blot, as represented in Fig. 6. The transferrin-targeted liposomal nanoparticles showed enhanced pEGFR and HuR protein downregulation upon treatment with the combination of gefitinib and HuR siRNA, compared with individual treatments.

## 6.3 LHRH Receptor

LHRH is a hormonal decapeptide that specifically targets LHRH receptors expressed on the cell membrane. Expression of LHRH is high in several cancers, including ovarian, endometrial, prostate, and breast cancer, whereas LHRH expression levels are very low in healthy tissues. Studies have shown that LHRH receptor-targeted dendrimers and liposomes produced enhanced accumulation of the drug at the tumor site (Minko et al., 2010; Saad et al., 2008).

Nukolva et al. developed targeted nanogels for delivery of chemotherapeutic agent cisplatin targeting luteinizing hormone receptors. They

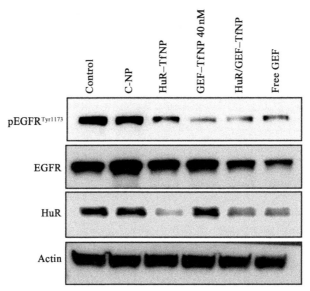

**Fig. 6** Western blot analysis of Tf receptor-targeted codelivery of HuR siRNA and gefitinib (40 nM) with DOTAP-chol liposome nanoparticles to HCC827 lung cancer cells. The results showed the reduction of HuR and pEGFR expression.

synthesized soft polymeric nanogels that can serve as carriers for hydrophilic drugs, such as cisplatin. These nanogels were less toxic and showed enhanced antitumor activity (Nukolova et al., 2013). Multifunctional lipid nanocarriers can deliver both chemotherapeutic drug and siRNA to the lungs through inhalation. These lipid nanocarriers were used for the delivery of anticancer drug PTX or DOX, and BCL2 siRNA. Inhalation resulted in the specific accumulation of these nanocarriers in the lung compared with other organs, whereas intravenous injection resulted in more accumulation in the liver, kidney, and spleen, and significantly less accumulation in the lung (Taratula, Kuzmov, Shah, Garbuzenko, & Minko, 2013). Iron oxide nanoparticles were coated with gold to form SPIONS-Au conjugated to the LHRH protein to target LHRH-overexpressing cancer cells. This study demonstrated that the SPIONs can generate a therapeutic amount of heat under magnetic fluid conditions, with controlled drug release (Mohammad & Al-Lohedan, 2017).

## 6.4 Folate Receptor

FR is membrane-bound protein on the cell surface, which strongly binds with folic acid molecules in neutral pH and transports them into cells. Folate mainly plays a role in DNA synthesis, methylation, and repair (Crider, Yang, Berry, & Bailey, 2012). It is widely overexpressed in many types of cancer tissues (Parker et al., 2005). Thus, FR can be used as a delivery target in cancer treatment by incorporating folic acid conjugation on the nanocarrier surfaces.

Fernandes et al. labeled chitosan with folic acid and used it as a carrier for siRNA delivery in HeLa and OV-3 cells, which express high levels of FRs. The researchers reported excellent siRNA transfection efficiency and low toxicity of the folate–chitosan nanoparticle in FR-positive cells (Fernandes et al., 2012). Dendrimer-based PAMAM polymer nanoparticles were covalently conjugated and used for hydrophobic baicalin (BAI) drug delivery. Administration of these PAMAM–FA/BAI complexes resulted in more cellular uptake in FR-positive HeLa cancer cells than in FR-negative A549 cells. Enhanced cell killing efficiency was observed with PAMAM–FA/BAI complexes compared with non-FA-modified PAMAM/BAI complexes (Lv et al., 2017). Recently, albumin coated on gold nanoparticles modified with FA- for FR-targeted drug delivery was combined with the chemotherapeutic DOX and computed tomography (CT) imaging. The Au-BSA-DOX-FA nanocomposites exhibited antitumor activity specifically on FR-overexpressing tumors, and no toxicity was observed in normal healthy tissues (Huang et al., 2017).

Our group recently published regarding FA-targeted HuR siRNA delivery in nonsmall cell lung cancer using DOTAP:cholesterol liposomes (HuR-FNP). FA was covalently conjugated on the surface of liposomes, and HuR siRNA was encapsulated on the liposomes through electrostatic interaction. HuR-FNP showed enhanced cell uptake and therapeutic effects in FRA-overexpressing H1299 cancer cells compared with normal CCD16 lung fibroblast cells with no FRA expression. HuR-FNP enhanced inhibition of cell migration in H1299 cells compared with C-FNP and control at both 24 and 48 h (Fig. 7), thus HuR-FNP exhibited activity on FR overexpressing tumors (Muralidharan et al., 2016).

## 6.5 Integrins

Integrins are cell adhesion molecules that comprise a large family of receptors involving alpha and beta subunits expressed on the cell surface. A peptide sequence with an RGD (arginine–glycone–aspartic acid) motif has an affinity toward integrins, such as $\alpha v\beta 3$, $\alpha v\beta 5$, and $\alpha 5\beta 1$, that are broadly expressed on the surface of tumor cells and tumor vasculature, but are largely

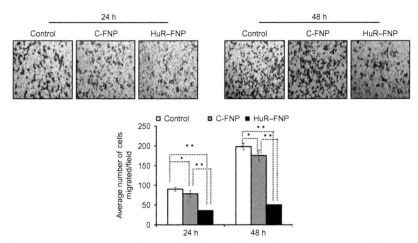

Fig. 7 Folic acid-conjugated DOTAP:cholesterol-based liposomes (HuR-FNP) greatly inhibit tumor cell migration compared with control siRNA (C-FNP) and untreated control at 24 and 48 h. *Figure reproduced from Muralidharan, R., Babu, A., Amreddy, N., Basalingappa, K., Mehta, M., Chen, A., ..., Ramesh, R. (2016). Folate receptor-targeted nanoparticle delivery of HuR-RNAi suppresses lung cancer cell proliferation and migration. Journal of Nanobiotechnology, 14(1), 47. doi:10.1186/s12951-016-0201-1. Creativecommons. org license 2016.*

absent from normal cells. Thus, RGD peptide is an attractive tumor-targeting protein for drug delivery using nanoparticles (Wang, Chui, & Ho, 2010). Two types of RGD peptides are commonly used as ligands to functionalize the nanoparticle surface: linear and cyclic. Cyclic RGD peptides [c(RGDfK)] showed high affinity and selectivity toward integrins compared with linear peptides, and thus can be used for tumor-targeted nanoparticle delivery of cancer therapeutics. Recently, Miura et al. utilized a micellar delivery system conjugated with cyclic RGD peptides for platinum drugs toward glioblastoma. Compared with nontargeted control peptide-conjugated micelles, cyclic RGD-conjugated micelle systems accumulated rapidly and had high permeability to the tumor parenchyma from vessels. The cyclic RGD-conjugated micelle delivery of platinum drug also resulted in significant tumor regression in an orthotopic mouse model of U87MG human glioblastoma (Miura et al., 2013).

Another derivative of RGD peptide, iRGD (CRGDKGPDC) peptide, has affinity toward integrin $\alpha v\beta 3$ and neuropilin (NRP) receptors overexpressed in cancer cells (Yin et al., 2017). In addition to targeting integrins, the iRGD peptide has tumor-homing properties that have been harnessed for tumor therapy and targeted drug delivery using nanoparticles

(Nie et al., 2014). In a typical study, iRGD peptide conjugated to paclitaxel albumin nanoparticles increased paclitaxel accumulation in tumor up to 4-folds when compared with unmodified nanoparticles (Sugahara et al., 2009). Another study utilized iRGD in conjunction with doxorubicin-conjugated gold nanoparticles–gelatin nanoparticle system (DOX–AuNPs–GNPs) to achieve deep tumor penetration and drug delivery (Cun et al., 2015). The use of iRGD increased the vascular and tumor permeability, resulting in enhanced tumor accumulation of nanoparticles and apoptosis induced by DOX.

## 7. CONCLUSION AND PROSPECTS

The main advantages of using delivery vehicles for cancer therapeutic agents are their enhanced bioavailability, controlled release, ability to prevent degradation, and preferential delivery to the tumor site. Therapeutic agents, such as chemotherapy drugs, nucleic acid therapeutics, small molecule inhibitors, and photodynamic agents can be loaded onto nanoparticles for cancer treatment. Each of these therapeutic agents differ in their physicochemical characteristics and their mechanisms of action, and require a specific dose to be released in the tumor site. However, many biological barriers must be overcome before drug reaches the target cancer cells.

Various nanoparticle formulations are currently being developed and explored as delivery vehicles for cancer therapeutics. These include a wide range of lipids, polymers of natural and synthetic origin, inorganic nanoparticles, and cellular vesicles that have been used for targeted delivery of biomolecules and therapeutic agents with promising in vitro and in vivo efficacy. Advanced delivery systems have also been designed to deliver drugs specifically to the tumor site, based on their stimuli-responsive properties. Targeted nanoparticles utilize ligands conjugated onto their surfaces to deliver therapeutic agents to specific receptors that are overexpressed in tumor cells, thereby enhancing their therapeutic efficiency. However, translating these nanoparticle systems to clinical settings is challenging. Obstacles, such as maintaining the stability of therapeutic agents, controlling their pharmacokinetics, reducing toxicity, and achieving targeted delivery, have been overcome to some extent by designing nanoparticles that can delay bioclearance, use of stimuli-responsive delivery strategy, and through a proper understanding of ligand–receptor interaction specific to tumor cells.

However, the unprecedented behavior of materials used for nanoparticle formulations, such as off-target effects or nonspecific toxicity, maintaining consistency in particle synthesis, and controlling penetration of biological barriers, are major hurdles to FDA approval. Therefore, many of the nanoparticle systems that appear promising in vitro may not be successful in vivo. Proper standards should be established for the examination of safety and efficacy issues before expanding the newly developed nanoparticle carriers into preclinical and clinical testing. Implementing proper regulatory measures, a deep understanding of tumor biology, and thoughtful use of technology advancements will speed the possible use of these nanoparticle systems in mainstream cancer treatment.

## ACKNOWLEDGMENTS

The study was supported in part by a grant received from the National Institutes of Health (NIH), R01 CA167516 (R.R.), an Institutional Development Award (IDeA) from the National Institute of General Medical Sciences (P20 GM103639) of the National Institutes of Health (A.M. and R.R.), and by funds received from the Stephenson Cancer Center Seed Grant (R.R.), Presbyterian Health Foundation Seed Grant (R.R.), Presbyterian Health Foundation Bridge Grant (R.R.), and Jim and Christy Everest Endowed Chair in Cancer Developmental Therapeutics (R.R.) at the University of Oklahoma Health Sciences Center. The authors thank Ms. Kathy Kyler at the office of the Vice President of Research, OUHSC, for editorial assistance. R.R. is an Oklahoma TSET Research Scholar and holds the Jim and Christy Everest Endowed Chair in Cancer Developmental Therapeutics.

## REFERENCES

Abbasi, E., Aval, S. F., Akbarzadeh, A., Milani, M., Nasrabadi, H. T., Joo, S. W., et al. (2014). Dendrimers: Synthesis, applications, and properties. *Nanoscale Research Letters*, *9*(1), 247. https://doi.org/10.1186/1556-276X-9-247.

Al-Ahmady, Z. S., Al-Jamal, W. T., Bossche, J. V., Bui, T. T., Drake, A. F., Mason, A. J., et al. (2012). Lipid–peptide vesicle nanoscale hybrids for triggered drug release by mild hyperthermia in vitro and in vivo. *ACS Nano*, *6*(10), 9335–9346. https://doi.org/10.1021/nn302148p.

Almeida, E., Bellettini, I. C., Garcia, F. P., Farinacio, M. T., Nakamura, C. V., Rubira, A. F., et al. (2017). Curcumin-loaded dual pH- and thermo-responsive magnetic microcarriers based on pectin maleate for drug delivery. *Carbohydrate Polymers*, *171*, 259–266. https://doi.org/10.1016/j.carbpol.2017.05.034.

Aluri, R., & Jayakannan, M. (2017). Development of l-tyrosine-based enzyme-responsive amphiphilic poly(ester-urethane) nanocarriers for multiple drug delivery to cancer cells. *Biomacromolecules*, *18*(1), 189–200. https://doi.org/10.1021/acs.biomac.6b01476.

Alvarez-Erviti, L., Seow, Y., Yin, H., Betts, C., Lakhal, S., & Wood, M. J. (2011). Delivery of siRNA to the mouse brain by systemic injection of targeted exosomes. *Nature Biotechnology*, *29*(4), 341–345. https://doi.org/10.1038/nbt.1807.

Alvarez-Lorenzo, C., Bromberg, L., & Concheiro, A. (2009). Light-sensitive intelligent drug delivery systems. *Photochemistry and Photobiology*, *85*(4), 848–860. https://doi.org/10.1111/j.1751-1097.2008.00530.x.

Amoli-Diva, M., Sadighi-Bonabi, R., & Pourghazi, K. (2017). Switchable on/off drug release from gold nanoparticles-grafted dual light- and temperature-responsive hydrogel for controlled drug delivery. *Materials Science & Engineering C, Materials for Biological Applications*, 76, 242–248. https://doi.org/10.1016/j.msec.2017.03.038.

Amreddy, N., Babu, A., Muralidharan, R., Munshi, A., & Ramesh, R. (2017). Polymeric nanoparticle-mediated gene delivery for lung cancer treatment. *Topics in Current Chemistry*, 375(2), 35. https://doi.org/10.1007/s41061-017-0128-5.

Andersen, M. O., Howard, K. A., Paludan, S. R., Besenbacher, F., & Kjems, J. (2008). Delivery of siRNA from lyophilized polymeric surfaces. *Biomaterials*, 29(4), 506–512. https://doi.org/10.1016/j.biomaterials.2007.10.003.

Arkin, M. R., & Wells, J. A. (2004). Small-molecule inhibitors of protein–protein interactions: Progressing towards the dream. *Nature Reviews. Drug Discovery*, 3(4), 301–317. https://doi.org/10.1038/nrd1343.

Ashton, S., Song, Y. H., Nolan, J., Cadogan, E., Murray, J., Odedra, R., et al. (2016). Aurora kinase inhibitor nanoparticles target tumors with favorable therapeutic index in vivo. *Science Translational Medicine*, 8(325), 325ra317. https://doi.org/10.1126/scitranslmed.aad2355.

Babu, A., Jeyasubramanian, K., Gunasekaran, P., & Murugesan, R. (2012). Gelatin nanocarrier enables efficient delivery and phototoxicity of hypocrellin B against a mice tumour model. *Journal of Biomedical Nanotechnology*, 8(1), 43–56.

Babu, A., & Ramesh, R. (2017). Multifaceted applications of chitosan in cancer drug delivery and therapy. *Marine Drugs*, 15(4), pii: E96. https://doi.org/10.3390/md15040096.

Babu, A., Wang, Q., Muralidharan, R., Shanker, M., Munshi, A., & Ramesh, R. (2014). Chitosan coated polylactic acid nanoparticle-mediated combinatorial delivery of cisplatin and siRNA/plasmid DNA chemosensitizes cisplatin-resistant human ovarian cancer cells. *Molecular Pharmaceutics*, 11(8), 2720–2733. https://doi.org/10.1021/mp500259e.

Baghbani, F., Chegeni, M., Moztarzadeh, F., Hadian-Ghazvini, S., & Raz, M. (2017). Novel ultrasound-responsive chitosan/perfluorohexane nanodroplets for image-guided smart delivery of an anticancer agent: Curcumin. *Materials Science & Engineering C: Materials for Biological Applications*, 74, 186–193. https://doi.org/10.1016/j.msec.2016.11.107.

Baghbani, F., & Moztarzadeh, F. (2017). Bypassing multidrug resistant ovarian cancer using ultrasound responsive doxorubicin/curcumin co-deliver alginate nanodroplets. *Colloids and Surfaces B: Biointerfaces*, 153, 132–140. https://doi.org/10.1016/j.colsurfb.2017.01.051.

Balazs, D. A., & Godbey, W. (2011). Liposomes for use in gene delivery. *Journal of Drug Delivery*, 2011, 326497. https://doi.org/10.1155/2011/326497.

Bernstein, E., Kim, S. Y., Carmell, M. A., Murchison, E. P., Alcorn, H., Li, M. Z., et al. (2003). Dicer is essential for mouse development. *Nature Genetics*, 35(3), 215–217. https://doi.org/10.1038/ng1253.

Bharti, C., Nagaich, U., Pal, A. K., & Gulati, N. (2015). Mesoporous silica nanoparticles in target drug delivery system: A review. *International Journal of Pharmaceutical Investigation*, 5(3), 124–133. https://doi.org/10.4103/2230-973X.160844.

Bhatt, R., de Vries, P., Tulinsky, J., Bellamy, G., Baker, B., Singer, J. W., et al. (2003). Synthesis and in vivo antitumor activity of poly(l-glutamic acid) conjugates of 20S-camptothecin. *Journal of Medicinal Chemistry*, 46(1), 190–193. https://doi.org/10.1021/jm020022r.

Bikram, M., & West, J. L. (2008). Thermo-responsive systems for controlled drug delivery. *Expert Opinion on Drug Delivery*, 5(10), 1077–1091. https://doi.org/10.1517/17425247.5.10.1077.

Brannon-Peppas, L., & Blanchette, J. O. (2004). Nanoparticle and targeted systems for cancer therapy. *Advanced Drug Delivery Reviews*, 56(11), 1649–1659. https://doi.org/10.1016/j.addr.2004.02.014.

Burrer, C. M., Auburn, H., Wang, X., Luo, J., Abulwerdi, F. A., Nikolovska-Coleska, Z., et al. (2017). Mcl-1 small-molecule inhibitors encapsulated into nanoparticles exhibit increased killing efficacy towards HCMV-infected monocytes. *Antiviral Research, 138*, 40–46. https://doi.org/10.1016/j.antiviral.2016.11.027.

Cai, W., Gao, T., Hong, H., & Sun, J. (2008). Applications of gold nanoparticles in cancer nanotechnology. *Nanotechnology, Science and Applications, 1*, 17–32. https://doi.org/10.2147/NSA.S3788.

Cai, Z., Zhang, H., Wei, Y., Wei, Y., Xie, Y., & Cong, F. (2017). Reduction- and pH-sensitive hyaluronan nanoparticles for delivery of iridium(III) anticancer drugs. *Biomacromolecules, 18*, 2102–2117. https://doi.org/10.1021/acs.biomac.7b00445.

Cancer Facts & Figures (2017). American Chemical Society, Retrieved from URL: https://www.cancer.org/research/cancer-facts-statistics/all-cancer-facts-figures/cancer-facts-figures-2017.html. on June 17, 2017.

Castano, A. P., Demidova, T. N., & Hamblin, M. R. (2004). Mechanisms in photodynamic therapy: Part one-photosensitizers, photochemistry and cellular localization. *Photodiagnosis and Photodynamic Therapy, 1*(4), 279–293. https://doi.org/10.1016/S1572-1000(05)00007-4.

Celikoglu, S. I., Karayel, T., Demirci, S., Celikoglu, F., & Cagatay, T. (1997). Direct injection of anti-cancer drugs into endobronchial tumours for palliation of major airway obstruction. *Postgraduate Medical Journal, 73*(857), 159–162.

Chatterjee, D. K., Fong, L. S., & Zhang, Y. (2008). Nanoparticles in photodynamic therapy: An emerging paradigm. *Advanced Drug Delivery Reviews, 60*(15), 1627–1637. https://doi.org/10.1016/j.addr.2008.08.003.

Chen, Y., Deng, X., Li, C., He, F., Liu, B., Hou, Z., et al. (2017). Stimuli-responsive nanocomposites for magnetic targeting synergistic multimodal therapy and T1/T2-weighted dual-mode imaging. *Nanomedicine, 13*(3), 875–883. https://doi.org/10.1016/j.nano.2016.12.004.

Chen, J., Ratnayaka, S., Alford, A., Kozlovskaya, V., Liu, F., Xue, B., et al. (2017). Theranostic multilayer capsules for ultrasound imaging and guided drug delivery. *ACS Nano, 11*(3), 3135–3146. https://doi.org/10.1021/acsnano.7b00151.

Chen, G., Wang, Y., Xie, R., & Gong, S. (2017). Tumor-targeted pH/redox dual-sensitive unimolecular nanoparticles for efficient siRNA delivery. *Journal of Controlled Release, 259*, 105–114. https://doi.org/10.1016/j.jconrel.2017.01.042.

Chen, Q., Xu, M., Zheng, W., Xu, T., Deng, H., & Liu, J. (2017). Se/Ru-decorated porous metal-organic framework nanoparticles for the delivery of pooled siRNAs to reversing multidrug resistance in Taxol-resistant breast cancer cells. *ACS Applied Materials & Interfaces, 9*(8), 6712–6724. https://doi.org/10.1021/acsami.6b12792.

Cheng, Y., Hao, J., Lee, L. A., Biewer, M. C., Wang, Q., & Stefan, M. C. (2012). Thermally controlled release of anticancer drug from self-assembled gamma-substituted amphiphilic poly(epsilon-caprolactone) micellar nanoparticles. *Biomacromolecules, 13*(7), 2163–2173. https://doi.org/10.1021/bm300823y.

Chenna, V., Hu, C., Pramanik, D., Aftab, B. T., Karikari, C., Campbell, N. R., et al. (2012). A polymeric nanoparticle encapsulated small-molecule inhibitor of hedgehog signaling (NanoHHI) bypasses secondary mutational resistance to smoothened antagonists. *Molecular Cancer Therapeutics, 11*(1), 165–173. https://doi.org/10.1158/1535-7163.MCT-11-0341.

Choi, S. H., Jin, S. E., Lee, M. K., Lim, S. J., Park, J. S., Kim, B. G., et al. (2008). Novel cationic solid lipid nanoparticles enhanced p53 gene transfer to lung cancer cells. *European Journal of Pharmaceutics and Biopharmaceutics, 68*(3), 545–554. https://doi.org/10.1016/j.ejpb.2007.07.011.

Choung, S., Kim, Y. J., Kim, S., Park, H. O., & Choi, Y. C. (2006). Chemical modification of siRNAs to improve serum stability without loss of efficacy. *Biochemical and Biophysical Research Communications, 342*(3), 919–927.

Crider, K. S., Yang, T. P., Berry, R. J., & Bailey, L. B. (2012). Folate and DNA methylation: A review of molecular mechanisms and the evidence for folate's role. *Advances in Nutrition, 3*(1), 21–38. https://doi.org/10.3945/an.111.000992.

Cross, D., & Burmester, J. K. (2006). Gene therapy for cancer treatment: Past, present and future. *Clinical Medicine & Research, 4*(3), 218–227.
Cun, X., Chen, J., Ruan, S., Zhang, L., Wan, J., He, Q., et al. (2015). A novel strategy through combining iRGD peptide with tumor-microenvironment-responsive and multistage nanoparticles for deep tumor penetration. *ACS Applied Materials & Interfaces, 7*(49), 27458–27466. https://doi.org/10.1021/acsami.5b09391.
Danafar, H., & Schumacher, U. (2016). MPEG–PCL copolymeric nanoparticles in drug delivery systems. *Cogent Medicine, 3,* Article No. 1142411.
Daniels, T. R., Bernabeu, E., Rodriguez, J. A., Patel, S., Kozman, M., Chiappetta, D. A., et al. (2012). The transferrin receptor and the targeted delivery of therapeutic agents against cancer. *Biochimica et Biophysica Acta, 1820*(3), 291–317. https://doi.org/10.1016/j.bbagen.2011.07.016.
de la Rica, R., Aili, D., & Stevens, M. M. (2012). Enzyme-responsive nanoparticles for drug release and diagnostics. *Advanced Drug Delivery Reviews, 64*(11), 967–978. https://doi.org/10.1016/j.addr.2012.01.002.
Deng, X., Cao, M., Zhang, J., Hu, K., Yin, Z., Zhou, Z., et al. (2014). Hyaluronic acid-chitosan nanoparticles for co-delivery of MiR-34a and doxorubicin in therapy against triple negative breast cancer. *Biomaterials, 35*(14), 4333–4344. https://doi.org/10.1016/j.biomaterials.2014.02.006.
Deng, L., Li, Q., Al-Rehili, S., Omar, H., Almalik, A., Alshamsan, A., et al. (2016). Hybrid iron oxide-graphene oxide-polysaccharides microcapsule: A micro-Matryoshka for on-demand drug release and antitumor therapy in vivo. *ACS Applied Materials & Interfaces, 8*(11), 6859–6868. https://doi.org/10.1021/acsami.6b00322.
Dufes, C., Uchegbu, I. F., & Schatzlein, A. G. (2005). Dendrimers in gene delivery. *Advanced Drug Delivery Reviews, 57*(15), 2177–2202. https://doi.org/10.1016/j.addr.2005.09.017.
Dumontet, C., & Sikic, B. I. (1999). Mechanisms of action of and resistance to antitubulin agents: Microtubule dynamics, drug transport, and cell death. *Journal of Clinical Oncology, 17*(3), 1061–1070. https://doi.org/10.1200/JCO.1999.17.3.1061.
Estrella, V., Chen, T., Lloyd, M., Wojtkowiak, J., Cornnell, H. H., Ibrahim-Hashim, A., et al. (2013). Acidity generated by the tumor microenvironment drives local invasion. *Cancer Research, 73*(5), 1524–1535. https://doi.org/10.1158/0008-5472.CAN-12-2796.
Fernandes, J. C., Qiu, X., Winnik, F. M., Benderdour, M., Zhang, X., Dai, K., et al. (2012). Low molecular weight chitosan conjugated with folate for siRNA delivery in vitro: Optimization studies. *International Journal of Nanomedicine, 7,* 5833–5845. https://doi.org/10.2147/IJN.S35567.
Ferrara, K. W. (2008). Driving delivery vehicles with ultrasound. *Advanced Drug Delivery Reviews, 60*(10), 1097–1102. https://doi.org/10.1016/j.addr.2008.03.002.
Fire, A., Xu, S., Montgomery, M. K., Kostas, S. A., Driver, S. E., & Mello, C. C. (1998). Potent and specific genetic interference by double-stranded RNA in Caenorhabditis elegans. *Nature, 391*(6669), 806–811. https://doi.org/10.1038/35888.
Flak, D., Yate, L., Nowaczyk, G., & Jurga, S. (2017). Hybrid ZnPc@$TiO_2$ nanostructures for targeted photodynamic therapy, bioimaging and doxorubicin delivery. *Materials Science & Engineering. C, Materials for Biological Applications, 78,* 1072–1085. https://doi.org/10.1016/j.msec.2017.04.107.
Fomina, N., Sankaranarayanan, J., & Almutairi, A. (2012). Photochemical mechanisms of light-triggered release from nanocarriers. *Advanced Drug Delivery Reviews, 64*(11), 1005–1020. https://doi.org/10.1016/j.addr.2012.02.006.
Fu, Y., & Kao, W. J. (2010). Drug release kinetics and transport mechanisms of non-degradable and degradable polymeric delivery systems. *Expert Opinion on Drug Delivery, 7*(4), 429–444. https://doi.org/10.1517/17425241003602259.
Giri, S., Trewyn, B. G., Stellmaker, M. P., & Lin, V. S. (2005). Stimuli-responsive controlled-release delivery system based on mesoporous silica nanorods capped with

magnetic nanoparticles. *Angewandte Chemie (International Ed. in English), 44*(32), 5038–5044. https://doi.org/10.1002/anie.200501819.

Gong, C., Qi, T., Wei, X., Qu, Y., Wu, Q., Luo, F., et al. (2013). Thermosensitive polymeric hydrogels as drug delivery systems. *Current Medicinal Chemistry, 20*(1), 79–94.

Gowda, R., Jones, N. R., Banerjee, S., & Robertson, G. P. (2013). Use of nanotechnology to develop multi-drug inhibitors for cancer therapy. *Journal of Nanomedicine & Nanotechnology, 4*(6), 184. https://doi.org/10.4172/2157-7439.1000184.

Gu, Y., Guo, Y., Wang, C., Xu, J., Wu, J., Kirk, T. B., et al. (2017). A polyamidoamne dendrimer functionalized graphene oxide for DOX and MMP-9 shRNA plasmid co-delivery. *Materials Science & Engineering C: Materials for Biological Applications, 70*(Pt. 1), 572–585. https://doi.org/10.1016/j.msec.2016.09.035.

Guerrero-Cazares, H., Tzeng, S. Y., Young, N. P., Abutaleb, A. O., Quinones-Hinojosa, A., & Green, J. J. (2014). Biodegradable polymeric nanoparticles show high efficacy and specificity at DNA delivery to human glioblastoma in vitro and in vivo. *ACS Nano, 8*(5), 5141–5153. https://doi.org/10.1021/nn501197v.

Haley, B., & Frenkel, E. (2008). Nanoparticles for drug delivery in cancer treatment. *Urologic Oncology, 26*(1), 57–64. https://doi.org/10.1016/j.urolonc.2007.03.015.

Han, Y., Zhang, P., Chen, Y., Sun, J., & Kong, F. (2014). Co-delivery of plasmid DNA and doxorubicin by solid lipid nanoparticles for lung cancer therapy. *International Journal of Molecular Medicine, 34*(1), 191–196. https://doi.org/10.3892/ijmm.2014.1770.

Harris, M., Ahmed, H., Barr, B., LeVine, D., Pace, L., Mohapatra, A., et al. (2017). Magnetic stimuli-responsive chitosan-based drug delivery biocomposite for multiple triggered release. *International Journal of Biological Macromolecules, 104*(Pt. B), 1407–1414. https://doi.org/10.1016/j.ijbiomac.2017.03.141.

Herman, T. S., Teicher, B. A., Jochelson, M., Clark, J., Svensson, G., & Coleman, C. N. (1988). Rationale for use of local hyperthermia with radiation therapy and selected anticancer drugs in locally advanced human malignancies. *International Journal of Hyperthermia, 4*(2), 143–158.

Hom, C., Lu, J., Liong, M., Luo, H., Li, Z., Zink, J. I., et al. (2010). Mesoporous silica nanoparticles facilitate delivery of siRNA to shutdown signaling pathways in mammalian cells. *Small, 6*(11), 1185–1190. https://doi.org/10.1002/smll.200901966.

Hsu, S. H., Yu, B., Wang, X., Lu, Y., Schmidt, C. R., Lee, R. J., et al. (2013). Cationic lipid nanoparticles for therapeutic delivery of siRNA and miRNA to murine liver tumor. *Nanomedicine, 9*(8), 1169–1180. https://doi.org/10.1016/j.nano.2013.05.007.

Hu, Q., Katti, P. S., & Gu, Z. (2014). Enzyme-responsive nanomaterials for controlled drug delivery. *Nanoscale, 6*(21), 12273–12286. https://doi.org/10.1039/c4nr04249b.

Hua, M. Y., Liu, H. L., Yang, H. W., Chen, P. Y., Tsai, R. Y., Huang, C. Y., et al. (2011). The effectiveness of a magnetic nanoparticle-based delivery system for BCNU in the treatment of gliomas. *Biomaterials, 32*(2), 516–527. https://doi.org/10.1016/j.biomaterials.2010.09.065.

Huang, H., Yang, D. P., Liu, M., Wang, X., Zhang, Z., Zhou, G., et al. (2017). pH-sensitive Au-BSA-DOX-FA nanocomposites for combined CT imaging and targeted drug delivery. *International Journal of Nanomedicine, 12*, 2829–2843. https://doi.org/10.2147/IJN.S128270.

Inicholson, R., WGee, J. M., & Harper, M. (2001). EGFR and cancer prognosis. *European Journal of Cancer, 37*, 9–15.

Isoldi, M. C., Visconti, M. A., & Castrucci, A. M. (2005). Anti-cancer drugs: Molecular mechanisms of action. *Mini Reviews in Medicinal Chemistry, 5*(7), 685–695.

Ito, I., Ji, L., Tanaka, F., Saito, Y., Gopalan, B., Branch, C. D., et al. (2004). Liposomal vector mediated delivery of the 3p FUS1 gene demonstrates potent antitumor activity against human lung cancer in vivo. *Cancer Gene Therapy, 11*(11), 733–739. https://doi.org/10.1038/sj.cgt.7700756.

Jiang, S., Eltoukhy, A. A., Love, K. T., Langer, R., & Anderson, D. G. (2013). Lipidoid-coated iron oxide nanoparticles for efficient DNA and siRNA delivery. *Nano Letters*, *13*(3), 1059–1064. https://doi.org/10.1021/nl304287a.

Jiang, L., Vader, P., & Schiffelers, R. M. (2017). Extracellular vesicles for nucleic acid delivery: Progress and prospects for safe RNA-based gene therapy. *Gene Therapy*, *24*(3), 157–166.

Johnsen, K. B., Gudbergsson, J. M., Skov, M. N., Pilgaard, L., Moos, T., & Duroux, M. (2014). A comprehensive overview of exosomes as drug delivery vehicles—Endogenous nanocarriers for targeted cancer therapy. *Biochimica et Biophysica Acta*, *1846*(1), 75–87.

Kasten, B. B., Liu, T., Nedrow-Byers, J. R., Benny, P. D., & Berkman, C. E. (2013). Targeting prostate cancer cells with PSMA inhibitor-guided gold nanoparticles. *Bioorganic & Medicinal Chemistry Letters*, *23*(2), 565–568. https://doi.org/10.1016/j.bmcl.2012.11.015.

Khan, M., Ong, Z. Y., Wiradharma, N., Attia, A. B., & Yang, Y. Y. (2012). Advanced materials for co-delivery of drugs and genes in cancer therapy. *Advanced Healthcare Materials*, *1*(4), 373–392. https://doi.org/10.1002/adhm.201200109.

Kim, S. I., Shin, D., Choi, T. H., Lee, J. C., Cheon, G. J., Kim, K. Y., et al. (2007). Systemic and specific delivery of small interfering RNAs to the liver mediated by apolipoprotein A–I. *Molecular Therapy*, *15*(6), 1145–1152. https://doi.org/10.1038/sj.mt.6300168.

Kneidl, B., Peller, M., Winter, G., Lindner, L. H., & Hossann, M. (2014). Thermosensitive liposomal drug delivery systems: State of the art review. *International Journal of Nanomedicine*, *9*, 4387–4398. https://doi.org/10.2147/IJN.S49297.

Kulsharova, G. K., Lee, M. B., Cheng, F., Haque, M., Choi, H., Kim, K., et al. (2013). In vitro and in vivo imaging of peptide-encapsulated polymer nanoparticles for cancer biomarker activated drug delivery. *IEEE Transactions on Nanobioscience*, *12*(4), 304–310. https://doi.org/10.1109/TNB.2013.2274781.

Kumar, B., Kulanthaivel, S., Mondal, A., Mishra, S., Banerjee, B., Bhaumik, A., et al. (2017). Mesoporous silica nanoparticle based enzyme responsive system for colon specific drug delivery through guar gum capping. *Colloids and Surfaces B: Biointerfaces*, *150*, 352–361. https://doi.org/10.1016/j.colsurfb.2016.10.049.

Kumar, C. S., & Mohammad, F. (2011). Magnetic nanomaterials for hyperthermia-based therapy and controlled drug delivery. *Advanced Drug Delivery Reviews*, *63*(9), 789–808. https://doi.org/10.1016/j.addr.2011.03.008.

Kundu, A. K., Iyer, S. V., Chandra, S., Adhikari, A. S., Iwakuma, T., & Mandal, T. K. (2017). Novel siRNA formulation to effectively knockdown mutant p53 in osteosarcoma. *PLoS One*, *12*(6), e0179168. https://doi.org/10.1371/journal.pone.0179168.

Landen, C. N., Jr., Chavez-Reyes, A., Bucana, C., Schmandt, R., Deavers, M. T., Lopez-Berestein, G., et al. (2005). Therapeutic EphA2 gene targeting in vivo using neutral liposomal small interfering RNA delivery. *Cancer Research*, *65*(15), 6910–6918.

Layek, B., Lipp, L., & Singh, J. (2015). Cell penetrating peptide conjugated chitosan for enhanced delivery of nucleic acid. *International Journal of Molecular Sciences*, *16*(12), 28912–28930. https://doi.org/10.3390/ijms161226142.

Lee, J. Y., Crake, C., Teo, B., Carugo, D., de Saint Victor, M., Seth, A., et al. (2017). Ultrasound-enhanced siRNA delivery using magnetic nanoparticle-loaded chitosan-deoxycholic acid nanodroplets. *Advanced Healthcare Materials*, *6*(8). https://doi.org/10.1002/adhm.201601246.

Lee, S. J., Yhee, J. Y., Kim, S. H., Kwon, I. C., & Kim, K. (2013). Biocompatible gelatin nanoparticles for tumor-targeted delivery of polymerized siRNA in tumor-bearing mice. *Journal of Controlled Release*, *172*(1), 358–366. https://doi.org/10.1016/j.jconrel.2013.09.002.

Lee, S. J., Yook, S., Yhee, J. Y., Yoon, H. Y., Kim, M. G., Ku, S. H., et al. (2015). Co-delivery of VEGF and Bcl-2 dual-targeted siRNA polymer using a single nanoparticle for synergistic anti-cancer effects in vivo. *Journal of Controlled Release*, *220*(Pt. B), 631–641. https://doi.org/10.1016/j.jconrel.2015.08.032.

Li, Y., Hu, H., Zhou, Q., Ao, Y., Xiao, C., Wan, J., et al. (2017). Alpha-amylase- and redox-responsive nanoparticles for tumor-targeted drug delivery. *ACS Applied Materials & Interfaces*, *9*(22), 19215–19230. https://doi.org/10.1021/acsami.7b04066.

Li, F., Li, T., Cao, W., Wang, L., & Xu, H. (2017). Near-infrared light stimuli-responsive synergistic therapy nanoplatforms based on the coordination of tellurium-containing block polymer and cisplatin for cancer treatment. *Biomaterials*, *133*, 208–218. https://doi.org/10.1016/j.biomaterials.2017.04.032.

Li, H., Wang, P., Deng, Y., Zeng, M., Tang, Y., Zhu, W. H., et al. (2017). Combination of active targeting, enzyme-triggered release and fluorescent dye into gold nanoclusters for endomicroscopy-guided photothermal/photodynamic therapy to pancreatic ductal adenocarcinoma. *Biomaterials*, *139*, 30–38. https://doi.org/10.1016/j.biomaterials.2017.05.030.

Liu, F. S. (2009). Mechanisms of chemotherapeutic drug resistance in cancer therapy—A quick review. *Taiwanese Journal of Obstetrics & Gynecology*, *48*(3), 239–244. https://doi.org/10.1016/S1028-4559(09)60296-5.

Liu, Y., Fu, S., Lin, L., Cao, Y., Xie, X., Yu, H., et al. (2017). Redox-sensitive pluronic F127-tocopherol micelles: Synthesis, characterization, and cytotoxicity evaluation. *International Journal of Nanomedicine*, *12*, 2635–2644. https://doi.org/10.2147/IJN.S122746.

Liu, S., Guo, Y., Huang, R., Li, J., Huang, S., Kuang, Y., et al. (2012). Gene and doxorubicin co-delivery system for targeting therapy of glioma. *Biomaterials*, *33*(19), 4907–4916. https://doi.org/10.1016/j.biomaterials.2012.03.031.

Liu, J., Huang, L., Tian, X., Chen, X., Shao, Y., Xie, F., et al. (2017). Magnetic and fluorescent Gd2O3:Yb+/Ln3+ nanoparticles for simultaneous upconversion luminescence/MR dual modal imaging and NIR-induced photodynamic therapy. *International Journal of Nanomedicine*, *12*, 1–14. https://doi.org/10.2147/IJN.S118938.

Lopes, J. R., Santos, G., Barata, P., Oliveira, R., & Lopes, C. M. (2013). Physical and chemical stimuli-responsive drug delivery systems: Targeted delivery and main routes of administration. *Current Pharmaceutical Design*, *19*(41), 7169–7184.

Lucero-Acuna, A., Jeffery, J. J., Abril, E. R., Nagle, R. B., Guzman, R., Pagel, M. D., et al. (2014). Nanoparticle delivery of an AKT/PDK1 inhibitor improves the therapeutic effect in pancreatic cancer. *International Journal of Nanomedicine*, *9*, 5653–5665. https://doi.org/10.2147/IJN.S68511.

Lungwitz, U., Breunig, M., Blunk, T., & Gopferich, A. (2005). Polyethylenimine-based non-viral gene delivery systems. *European Journal of Pharmaceutics and Biopharmaceutics*, *60*(2), 247–266. https://doi.org/10.1016/j.ejpb.2004.11.011.

Lv, T., Yu, T., Fang, Y., Zhang, S., Jiang, M., Zhang, H., et al. (2017). Role of generation on folic acid-modified poly(amidoamine) dendrimers for targeted delivery of baicalin to cancer cells. *Materials Science & Engineering C: Materials for Biological Applications*, *75*, 182–190. https://doi.org/10.1016/j.msec.2016.12.134.

Madaan, K., Kumar, S., Poonia, N., Lather, V., & Pandita, D. (2014). Dendrimers in drug delivery and targeting: Drug–dendrimer interactions and toxicity issues. *Journal of Pharmacy & Bioallied Sciences*, *6*(3), 139–150. https://doi.org/10.4103/0975-7406.130965.

Madani, S. Y., Naderi, N., Dissanayake, O., Tan, A., & Seifalian, A. M. (2011). A new era of cancer treatment: Carbon nanotubes as drug delivery tools. *International Journal of Nanomedicine*, *6*, 2963–2979. https://doi.org/10.2147/IJN.S16923.

McBain, S. C., Yiu, H. H., & Dobson, J. (2008). Magnetic nanoparticles for gene and drug delivery. *International Journal of Nanomedicine*, *3*(2), 169–180.

Medarova, Z., Balcioglu, M., & Yigit, M. V. (2016). Controlling RNA expression in cancer using iron oxide nanoparticles detectable by MRI and in vivo optical imaging. *Methods in Molecular Biology*, *1372*, 163–179. https://doi.org/10.1007/978-1-4939-3148-4_13.

Mendes, R., Fernandes, A. R., & Baptista, P. V. (2017). Gold nanoparticle approach to the selective delivery of gene silencing in cancer-the case for combined delivery? *Genes (Basel)*, *8*(3), pii: E94. https://doi.org/10.3390/genes8030094.

Minko, T., Patil, M. L., Zhang, M., Khandare, J. J., Saad, M., Chandna, P., et al. (2010). LHRH-targeted nanoparticles for cancer therapeutics. *Methods in Molecular Biology*, *624*, 281–294. https://doi.org/10.1007/978-1-60761-609-2_19.

Miura, Y., Takenaka, T., Toh, K., Wu, S., Nishihara, H., Kano, M. R., et al. (2013). Cyclic RGD-linked polymeric micelles for targeted delivery of platinum anticancer drugs to glioblastoma through the blood–brain tumor barrier. *ACS Nano*, *7*(10), 8583–8592. https://doi.org/10.1021/nn402662d.

Mohammad, F., & Al-Lohedan, H. A. (2017). Luteinizing hormone-releasing hormone targeted superparamagnetic gold nanoshells for a combination therapy of hyperthermia and controlled drug delivery. *Materials Science & Engineering C: Materials for Biological Applications*, *76*, 692–700. https://doi.org/10.1016/j.msec.2017.03.162.

Monks, A., Scudiero, D., Skehan, P., Shoemaker, R., Paull, K., Vistica, D., et al. (1991). Feasibility of a high-flux anticancer drug screen using a diverse panel of cultured human tumor cell lines. *Journal of the National Cancer Institute*, *83*(11), 757–766.

Morabito, A., Piccirillo, M. C., Monaco, K., Pacilio, C., Nuzzo, F., Chiodini, P., et al. (2007). First-line chemotherapy for HER-2 negative metastatic breast cancer patients who received anthracyclines as adjuvant treatment. *The Oncologist*, *12*(11), 1288–1298. https://doi.org/10.1634/theoncologist.12-11-1288.

Morrow, G. R. (1985). The effect of a susceptibility to motion sickness on the side effects of cancer chemotherapy. *Cancer*, *55*(12), 2766–2770.

Muller, R. H., Mader, K., & Gohla, S. (2000). Solid lipid nanoparticles (SLN) for controlled drug delivery—A review of the state of the art. *European Journal of Pharmaceutics and Biopharmaceutics*, *50*(1), 161–177.

Muralidharan, R., Babu, A., Amreddy, N., Basalingappa, K., Mehta, M., Chen, A., et al. (2016). Folate receptor-targeted nanoparticle delivery of HuR-RNAi suppresses lung cancer cell proliferation and migration. *Journal of Nanobiotechnology*, *14*(1), 47. https://doi.org/10.1186/s12951-016-0201-1.

Muralidharan, R., Babu, A., Amreddy, N., Srivastava, A., Chen, A., Zhao, Y. D., et al. (2017). Tumor-targeted nanoparticle delivery of HuR siRNA inhibits lung tumor growth in vitro and in vivo by disrupting the oncogenic activity of the RNA-binding protein HuR. *Molecular Cancer Therapeutics*, *16*(8), 1470–1486. https://doi.org/10.1158/1535-7163.MCT-17-0134.

Nie, X., Zhang, J., Xu, Q., Liu, X., Li, Y., Wu, Y., et al. (2014). Targeting peptide iRGD-conjugated amphiphilic chitosan-co-PLA/DPPE drug delivery system for enhanced tumor therapy. *Journal of Materials Chemistry B*, *2*, 3232–3242.

Nukolova, N. V., Oberoi, H. S., Zhao, Y., Chekhonin, V. P., Kabanov, A. V., & Bronich, T. K. (2013). LHRH-targeted nanogels as a delivery system for cisplatin to ovarian cancer. *Molecular Pharmaceutics*, *10*(10), 3913–3921. https://doi.org/10.1021/mp4003688.

Okayama, H., Kohno, T., Ishii, Y., Shimada, Y., Shiraishi, K., Iwakawa, R., et al. (2012). Identification of genes upregulated in ALK-positive and EGFR/KRAS/ALK-negative lung adenocarcinomas. *Cancer Research*, *72*(1), 100–111. https://doi.org/10.1158/0008-5472.CAN-11-1403.

Palchaudhuri, R., & Hergenrother, P. J. (2007). DNA as a target for anticancer compounds: Methods to determine the mode of binding and the mechanism of action. *Current Opinion in Biotechnology*, *18*(6), 497–503. https://doi.org/10.1016/j.copbio.2007.09.006.

Paliwal, S., & Mitragotri, S. (2006). Ultrasound-induced cavitation: Applications in drug and gene delivery. *Expert Opinion on Drug Delivery*, *3*(6), 713–726. https://doi.org/10.1517/17425247.3.6.713.

Panneerselvam, J., Jin, J., Shanker, M., Lauderdale, J., Bates, J., Wang, Q., et al. (2015). IL-24 inhibits lung cancer cell migration and invasion by disrupting the SDF-1/CXCR4 signaling axis. *PLoS One, 10*(3), e0122439. https://doi.org/10.1371/journal.pone.0122439.

Panneerselvam, J., Srivastava, A., Muralidharan, R., Wang, Q., Zheng, W., Zhao, L., et al. (2016). IL-24 modulates the high mobility group (HMG) A1/miR222/AKT signaling in lung cancer cells. *OncoTarget, 7*(43), 70247–70263. https://doi.org/10.18632/oncotarget.11838.

Panyam, J., & Labhasetwar, V. (2003). Biodegradable nanoparticles for drug and gene delivery to cells and tissue. *Advanced Drug Delivery Reviews, 55*(3), 329–347.

Parhi, P., Mohanty, C., & Sahoo, S. K. (2012). Nanotechnology-based combinational drug delivery: An emerging approach for cancer therapy. *Drug Discovery Today, 17*(17–18), 1044–1052. https://doi.org/10.1016/j.drudis.2012.05.010.

Parker, N., Turk, M. J., Westrick, E., Lewis, J. D., Low, P. S., & Leamon, C. P. (2005). Folate receptor expression in carcinomas and normal tissues determined by a quantitative radioligand binding assay. *Analytical Biochemistry, 338*(2), 284–293. https://doi.org/10.1016/j.ab.2004.12.026.

Patel, M. P., Patel, R. R., & Patel, J. K. (2010). Chitosan mediated targeted drug delivery system: A review. *Journal of Pharmacy & Pharmaceutical Sciences, 13*(4), 536–557.

Patil, S. D., Rhodes, D. G., & Burgess, D. J. (2004). Anionic liposomal delivery system for DNA transfection. *The AAPS Journal, 6*(4), e29. https://doi.org/10.1208/aapsj060429.

Pencheva, N., Tran, H., Buss, C., Huh, D., Drobnjak, M., Busam, K., et al. (2012). Convergent multi-miRNA targeting of ApoE drives LRP1/LRP8-dependent melanoma metastasis and angiogenesis. *Cell, 151*(5), 1068–1082. https://doi.org/10.1016/j.cell.2012.10.028.

Perry, J., Chambers, A., Spithoff, K., & Laperriere, N. (2007). Gliadel wafers in the treatment of malignant glioma: A systematic review. *Current Oncology, 14*(5), 189–194.

Perumal, V., Banerjee, S., Das, S., Sen, R. K., & Mandal, M. (2011). Effect of liposomal celecoxib on proliferation of colon cancer cell and inhibition of DMBA-induced tumor in rat model. *Cancer Nanotechnology, 2*(1–6), 67–79. https://doi.org/10.1007/s12645-011-0017-5.

Prabaharan, M., Grailer, J. J., Pilla, S., Steeber, D. A., & Gong, S. (2009). Amphiphilic multi-arm-block copolymer conjugated with doxorubicin via pH-sensitive hydrazone bond for tumor-targeted drug delivery. *Biomaterials, 30*(29), 5757–5766. https://doi.org/10.1016/j.biomaterials.2009.07.020.

Prabhu, S., Uzzaman, V., & Guruvayoorappan, G. C. (2011). Nanoparticles in drug delivery and cancer therapy: The Giant rats tail. *Journal of Cancer Therapy, 2*, 325–334.

Puoci, F., Iemma, F., & Picci, N. (2008). Stimuli-responsive molecularly imprinted polymers for drug delivery: A review. *Current Drug Delivery, 5*(2), 85–96.

Qiao, H., Zhu, Z., Fang, D., Sun, Y., Kang, C., Di, L., et al. (2017). Redox-triggered mitoxantrone prodrug micelles for overcoming multidrug-resistant breast cancer. *Journal of Drug Targeting, 18*, 1–11. https://doi.org/10.1080/1061186X.2017.1339195.

Qiu, L., Zhu, M., Gong, K., Peng, H., Ge, L., Zhao, L., et al. (2017). pH-triggered degradable polymeric micelles for targeted anti-tumor drug delivery. *Materials Science & Engineering C: Materials for Biological Applications, 78*, 912–922. https://doi.org/10.1016/j.msec.2017.04.137.

Saad, M., Garbuzenko, O. B., Ber, E., Chandna, P., Khandare, J. J., Pozharov, V. P., et al. (2008). Receptor targeted polymers, dendrimers, liposomes: Which nanocarrier is the most efficient for tumor-specific treatment and imaging? *Journal of Controlled Release, 130*(2), 107–114. https://doi.org/10.1016/j.jconrel.2008.05.024.

Saavedra-Alonso, S., Zapata-Benavides, P., Chavez-Escamilla, A. K., Manilla-Munoz, E., Zamora-Avila, D. E., Franco-Molina, M. A., et al. (2016). WT1 shRNA delivery using transferrin-conjugated PEG liposomes in an in vivo model of melanoma. *Experimental and Therapeutic Medicine, 12*(6), 3778–3784. https://doi.org/10.3892/etm.2016.3851.

Sabbatini, P., Aghajanian, C., Dizon, D., Anderson, S., Dupont, J., Brown, J. V., et al. (2004). Phase II study of CT-2103 in patients with recurrent epithelial ovarian, fallopian tube, or primary peritoneal carcinoma. *Journal of Clinical Oncology, 22*(22), 4523–4531. https://doi.org/10.1200/JCO.2004.12.043.

Sahoo, S. K., Ma, W., & Labhasetwar, V. (2004). Efficacy of transferrin-conjugated paclitaxel-loaded nanoparticles in a murine model of prostate cancer. *International Journal of Cancer, 112*(2), 335–340. https://doi.org/10.1002/ijc.20405.

Sardo, C., Bassi, B., Craparo, E. F., Scialabba, C., Cabrini, E., Dacarro, G., et al. (2017). Gold nanostar-polymer hybrids for siRNA delivery: Polymer design towards colloidal stability and in vitro studies on breast cancer cells. *International Journal of Pharmaceutics, 519*(1–2), 113–124. https://doi.org/10.1016/j.ijpharm.2017.01.022.

Shahabipour, F., Barati, N., Johnston, T. P., Derosa, G., Maffioli, P., & Sahebkar, A. (2017). Exosomes: Nanoparticulate tools for RNA interference and drug delivery. *Journal of Cellular Physiology, 232*(7), 1660–1668.

Shih, Y. H., Luo, T. Y., Chiang, P. F., Yao, C. J., Lin, W. J., Peng, C. L., et al. (2017). EGFR-targeted micelles containing near-infrared dye for enhanced photothermal therapy in colorectal cancer. *Journal of Controlled Release, 258*, 196–207. https://doi.org/10.1016/j.jconrel.2017.04.031.

Shu, D., Li, H., Shu, Y., Xiong, G., Carson, W. E., 3rd, Haque, F., et al. (2015). Systemic delivery of anti-miRNA for suppression of triple negative breast cancer utilizing RNA nanotechnology. *ACS Nano, 9*(10), 9731–9740. https://doi.org/10.1021/acsnano.5b02471.

Simoes, S., Filipe, A., Faneca, H., Mano, M., Penacho, N., Duzgunes, N., et al. (2005). Cationic liposomes for gene delivery. *Expert Opinion on Drug Delivery, 2*(2), 237–254. https://doi.org/10.1517/17425247.2.2.237.

Song, C. Z. (2007). Gene silencing therapy against cancer. In K. K. Hunt, S. A. Vorburger, & S. G. Swisher (Eds.), *Gene therapy for cancer* (pp. 185–196). Human Press.

Srivastava, A., Amreddy, N., Babu, A., Panneerselvam, J., Mehta, M., Muralidharan, R., et al. (2016). Nanosomes carrying doxorubicin exhibit potent anticancer activity against human lung cancer cells. *Scientific Reports, 6*, 38541. https://doi.org/10.1038/srep38541.

Srivastava, A., Babu, A., Filant, J., Moxley, K. M., Ruskin, R., Dhanasekaran, D., et al. (2016). Exploitation of exosomes as nanocarriers for gene-, chemo-, and immunetherapy of cancer. *Journal of Biomedical Nanotechnology, 12*(6), 1159–1173.

Srivastava, A., Filant, J., Moxley, K. M., Sood, A., McMeekin, S., & Ramesh, R. (2015). Exosomes: A role for naturally occurring nanovesicles in cancer growth, diagnosis and treatment. *Current Gene Therapy, 15*(2), 182–192.

Subramanian, N., Kanwar, J. R., Kanwar, R. K., & Krishnakumar, S. (2015). Blocking the maturation of OncomiRNAs using pri-miRNA-17 approximately 92 aptamer in retinoblastoma. *Nucleic Acid Therapeutics, 25*(1), 47–52. https://doi.org/10.1089/nat.2014.0507.

Sugahara, K. N., Teesalu, T., Karmali, P. P., Kotamraju, V. R., Agemy, L., Girard, O. M., et al. (2009). Tissue-penetrating delivery of compounds and nanoparticles into tumors. *Cancer Cell, 16*(6), 510–520. https://doi.org/10.1016/j.ccr.2009.10.013.

Sun, H., Meng, F., Cheng, R., Deng, C., & Zhong, Z. (2014). Reduction-responsive polymeric micelles and vesicles for triggered intracellular drug release. *Antioxidants & Redox Signaling, 21*(5), 755–767. https://doi.org/10.1089/ars.2013.5733.

Szwed, M., Kania, K. D., & Jozwiak, Z. (2014). Relationship between therapeutic efficacy of doxorubicin-transferrin conjugate and expression of P-glycoprotein in chronic erythromyeloblastoid leukemia cells sensitive and resistant to doxorubicin. *Cellular Oncology (Dordrecht), 37*(6), 421–428. https://doi.org/10.1007/s13402-014-0205-5.

Tagami, T., Foltz, W. D., Ernsting, M. J., Lee, C. M., Tannock, I. F., May, J. P., et al. (2011). MRI monitoring of intratumoral drug delivery and prediction of the therapeutic effect with a multifunctional thermosensitive liposome. *Biomaterials, 32*(27), 6570–6578. https://doi.org/10.1016/j.biomaterials.2011.05.029.

Taratula, O., Kuzmov, A., Shah, M., Garbuzenko, O. B., & Minko, T. (2013). Nanostructured lipid carriers as multifunctional nanomedicine platform for pulmonary co-delivery of anticancer drugs and siRNA. *Journal of Controlled Release, 171*(3), 349–357. https://doi.org/10.1016/j.jconrel.2013.04.018.

Tatiparti, K., Sau, S., Kashaw, S. K., & Iyer, A. K. (2017). siRNA delivery strategies: A comprehensive review of recent developments. *Nanomaterials (Basel), 7*(4), pii: E77. https://doi.org/10.3390/nano7040077.

Telli, M. L., & Carlson, R. W. (2009). First-line chemotherapy for metastatic breast cancer. *Clinical Breast Cancer, 2*, S66–72.

Usuda, J., Tsutsui, H., Honda, H., Ichinose, S., Ishizumi, T., Hirata, T., et al. (2007). Photodynamic therapy for lung cancers based on novel photodynamic diagnosis using talaporfin sodium (NPe6) and autofluorescence bronchoscopy. *Lung Cancer, 58*(3), 317–323. https://doi.org/10.1016/j.lungcan.2007.06.026.

Vankayala, R., Huang, Y. K., Kalluru, P., Chiang, C. S., & Hwang, K. C. (2014). First demonstration of gold nanorods-mediated photodynamic therapeutic destruction of tumors via near infra-red light activation. *Small, 10*(8), 1612–1622. https://doi.org/10.1002/smll.201302719.

Veber, D. F., Johnson, S. R., Cheng, H. Y., Smith, B. R., Ward, K. W., & Kopple, K. D. (2002). Molecular properties that influence the oral bioavailability of drug candidates. *Journal of Medicinal Chemistry, 45*(12), 2615–2623.

Wang, T. Y., Choe, J. W., Pu, K., Devulapally, R., Bachawal, S., Machtaler, S., et al. (2015). Ultrasound-guided delivery of microRNA loaded nanoparticles into cancer. *Journal of Controlled Release, 203*, 99–108. https://doi.org/10.1016/j.jconrel.2015.02.018.

Wang, Z., Chui, W. K., & Ho, P. C. (2010). Integrin targeted drug and gene delivery. *Expert Opinion on Drug Delivery, 7*(2), 159–171. https://doi.org/10.1517/17425240903468696.

Wang, M., Geilich, B. M., Keidar, M., & Webster, T. J. (2017). Killing malignant melanoma cells with protoporphyrin IX-loaded polymersome-mediated photodynamic therapy and cold atmospheric plasma. *International Journal of Nanomedicine, 12*, 4117–4127. https://doi.org/10.2147/IJN.S129266.

Wang, K., Kievit, F. M., Sham, J. G., Jeon, M., Stephen, Z. R., Bakthavatsalam, A., et al. (2016). Iron-oxide-based nanovector for tumor targeted siRNA delivery in an orthotopic hepatocellular carcinoma xenograft mouse model. *Small, 12*(4), 477–487. https://doi.org/10.1002/smll.201501985.

Wang, F., Li, C., Cheng, J., & Yuan, Z. (2016). Recent advances on inorganic nanoparticle-based cancer therapeutic agents. *International Journal of Environmental Research and Public Health, 13*(12), 1182. https://doi.org/10.3390/ijerph13121182.

Wang, J., Lu, Z., Wientjes, M. G., & Au, J. L. (2010). Delivery of siRNA therapeutics: Barriers and carriers. *The AAPS Journal, 12*(4), 492–503. https://doi.org/10.1208/s12248-010-9210-4.

Wang, Z., Qiao, R., Tang, N., Lu, Z., Wang, H., Zhang, Z., et al. (2017). Active targeting theranostic iron oxide nanoparticles for MRI and magnetic resonance-guided focused ultrasound ablation of lung cancer. *Biomaterials, 127*, 25–35. https://doi.org/10.1016/j.biomaterials.2017.02.037.

Wang, H., Tam, Y. Y., Chen, S., Zaifman, J., van der Meel, R., Ciufolini, M. A., et al. (2016). The Niemann–Pick C1 inhibitor NP3.47 enhances gene silencing potency of lipid nanoparticles containing siRNA. *Molecular Therapy, 24*(12), 2100–2108. https://doi.org/10.1038/mt.2016.179.

Wang, C., Tao, H., Cheng, L., & Liu, Z. (2011). Near-infrared light induced in vivo photodynamic therapy of cancer based on upconversion nanoparticles. *Biomaterials*, *32*(26), 6145–6154. https://doi.org/10.1016/j.biomaterials.2011.05.007.

Wang, Y., Yu, L., Han, L., Sha, X., & Fang, X. (2007). Difunctional pluronic copolymer micelles for paclitaxel delivery: Synergistic effect of folate-mediated targeting and pluronic-mediated overcoming multidrug resistance in tumor cell lines. *International Journal of Pharmaceutics*, *337*(1–2), 63–73. https://doi.org/10.1016/j.ijpharm.2006.12.033.

Wang, Z. Y., Zhang, H., Yang, Y., Xie, X. Y., Yang, Y. F., Li, Z., et al. (2016). Preparation, characterization, and efficacy of thermosensitive liposomes containing paclitaxel. *Drug Delivery*, *23*(4), 1222–1231. https://doi.org/10.3109/10717544.2015.1122674.

Wang, H., Zhao, P., Su, W., Wang, S., Liao, Z., Niu, R., et al. (2010). PLGA/polymeric liposome for targeted drug and gene co-delivery. *Biomaterials*, *31*(33), 8741–8748. https://doi.org/10.1016/j.biomaterials.2010.07.082.

Watts, J. K., & Corey, D. R. (2012). Silencing disease genes in the laboratory and the clinic. *The Journal of Pathology*, *226*(2), 365–379. https://doi.org/10.1002/path.2993.

Weiss, W. A., Taylor, S. S., & Shokat, K. M. (2007). Recognizing and exploiting differences between RNAi and small-molecule inhibitors. *Nature Chemical Biology*, *3*(12), 739–744. https://doi.org/10.1038/nchembio1207-739.

Wen, H. Y., Dong, H. Q., Xie, W. J., Li, Y. Y., Wang, K., Pauletti, G. M., et al. (2011). Rapidly disassembling nanomicelles with disulfide-linked PEG shells for glutathione-mediated intracellular drug delivery. *Chemical Communications*, *47*(12), 3550–3552. https://doi.org/10.1039/c0cc04983b.

Wu, Y., Wang, W., Chen, Y., Huang, K., Shuai, X., Chen, Q., et al. (2010). The investigation of polymer-siRNA nanoparticle for gene therapy of gastric cancer in vitro. *International Journal of Nanomedicine*, *5*, 129–136.

www.clinicaltrial.gov website accessed on June 28, 2017.

Xie, Y., Killinger, B., Moszczynska, A., & Merkel, O. M. (2016). Targeted delivery of siRNA to transferrin receptor overexpressing tumor cells via peptide modified polyethylenimine. *Molecules*, *21*(10), pii: E1334. https://doi.org/10.3390/molecules21101334.

Xie, Y., Kim, N. H., Nadithe, V., Schalk, D., Thakur, A., Kilic, A., et al. (2016). Targeted delivery of siRNA to activated T cells via transferrin-polyethylenimine (Tf-PEI) as a potential therapy of asthma. *Journal of Controlled Release*, *229*, 120–129. https://doi.org/10.1016/j.jconrel.2016.03.029.

Xu, J., Singh, A., & Amiji, M. M. (2014). Redox-responsive targeted gelatin nanoparticles for delivery of combination wt-p53 expressing plasmid DNA and gemcitabine in the treatment of pancreatic cancer. *BMC Cancer*, *14*, 75. https://doi.org/10.1186/1471-2407-14-75.

Xu, Q., Xia, Y., Wang, C. H., & Pack, D. W. (2012). Monodisperse double-walled microspheres loaded with chitosan-p53 nanoparticles and doxorubicin for combined gene therapy and chemotherapy. *Journal of Controlled Release*, *163*(2), 130–135. https://doi.org/10.1016/j.jconrel.2012.08.032.

Yin, T., Wang, L., Yin, L., Zhou, J., & Huo, M. (2015). Co-delivery of hydrophobic paclitaxel and hydrophilic AURKA specific siRNA by redox-sensitive micelles for effective treatment of breast cancer. *Biomaterials*, *61*, 10–25. https://doi.org/10.1016/j.biomaterials.2015.05.022.

Yin, H., Yang, J., Zhang, Q., Yang, J., Wang, H., Xu, J., et al. (2017). iRGD as a tumorpenetrating peptide for cancer therapy (Review). *Molecular Medicine Reports*, *15*(5), 2925–2930. https://doi.org/10.3892/mmr.2017.6419.

Yokoyama, T., Tam, J., Kuroda, S., Scott, A. W., Aaron, J., Larson, T., et al. (2011). EGFR-targeted hybrid plasmonic magnetic nanoparticles synergistically induce autophagy and apoptosis in non-small cell lung cancer cells. *PLoS One*, *6*(11), e25507. https://doi.org/10.1371/journal.pone.0025507.

Yoo, S. S., Razzak, R., Bedard, E., Guo, L., Shaw, A. R., Moore, R. B., et al. (2014). Layered gadolinium-based nanoparticle as a novel delivery platform for microRNA therapeutics. *Nanotechnology, 25*(42), 425102. https://doi.org/10.1088/0957-4484/25/42/425102.

Yoshida, T., Tokashiki, R., Ito, H., Shimizu, A., Nakamura, K., Hiramatsu, H., et al. (2008). Therapeutic effects of a new photosensitizer for photodynamic therapy of early head and neck cancer in relation to tissue concentration. *Auris Nasus Larynx, 35*(4), 545–551. https://doi.org/10.1016/j.anl.2007.10.008.

Younes, R. N., Pereira, J. R., Fares, A. L., & Gross, J. L. (2011). Chemotherapy beyond first-line in stage IV metastatic non-small cell lung cancer. *Revista Da Associacao Medica Brasileira, 57*(6), 686–691.

Yu, X., Trase, I., Ren, M., Duval, K., Guo, X., & Chen, Z. (2016). Design of nanoparticle-based carriers for targeted drug delivery. *Journal of Nanomaterials, 2016*, pii: 1087250. https://doi.org/10.1155/2016/1087250.

Zamore, P. D., Tuschl, T., Sharp, P. A., & Bartel, D. P. (2000). RNAi: Double-stranded RNA directs the ATP-dependent cleavage of mRNA at 21 to 23 nucleotide intervals. *Cell, 101*(1), 25–33. https://doi.org/10.1016/S0092-8674(00)80620-0.

Zhang, J. L., Gong, J. H., Xing, L., Cui, P. F., Qiao, J. B., He, Y. J., et al. (2016). One-step assembly of polymeric demethylcantharate prodrug/Akt1 shRNA complexes for enhanced cancer therapy. *International Journal of Pharmaceutics, 513*(1–2), 612–627. https://doi.org/10.1016/j.ijpharm.2016.09.070.

Zhang, X., Liu, N., Shao, Z., Qiu, H., Yao, H., Ji, J., et al. (2017). Folate-targeted nanoparticle delivery of androgen receptor shRNA enhances the sensitivity of hormone-independent prostate cancer to radiotherapy. *Nanomedicine, 13*(4), 1309–1321. https://doi.org/10.1016/j.nano.2017.01.015.

Zhang, C., Pan, D., Li, J., Hu, J., Bains, A., Guys, N., et al. (2017). Enzyme-responsive peptide dendrimer-gemcitabine conjugate as a controlled-release drug delivery vehicle with enhanced antitumor efficacy. *Acta Biomaterialia, 55*, 153–162. https://doi.org/10.1016/j.actbio.2017.02.047.

Zhang, J., Qian, Z., & Gu, Y. (2009). In vivo anti-tumor efficacy of docetaxel-loaded thermally responsive nanohydrogel. *Nanotechnology, 20*(32). 325102. https://doi.org/10.1088/0957-4484/20/32/325102.

Zhang, J., & Saltzman, M. (2013). Engineering biodegradable nanoparticles for drug and gene delivery. *Chemical Engineering Progress, 109*(3), 25–30.

Zhang, L., Wang, T., Yang, L., Liu, C., Wang, C., Liu, H., et al. (2012). General route to multifunctional uniform yolk/mesoporous silica shell nanocapsules: A platform for simultaneous cancer-targeted imaging and magnetically guided drug delivery. *Chemistry, 18*(39), 12512–12521. https://doi.org/10.1002/chem.201200030.

Zhang, H., Zhang, H., Zhu, X., Zhang, X., Chen, Q., Chen, J., et al. (2017). Visible-light-sensitive titanium dioxide nanoplatform for tumor-responsive $Fe^{2+}$ liberating and artemisinin delivery. *OncoTarget, 8*(35), 58738–58753. https://doi.org/10.18632/oncotarget.17639.

Zhao, Y., & Huang, L. (2014). Lipid nanoparticles for gene delivery. *Advances in Genetics, 88*, 13–36. https://doi.org/10.1016/B978-0-12-800148-6.00002-X.

Zhao, X., Li, F., Li, Y., Wang, H., Ren, H., Chen, J., et al. (2015). Co-delivery of HIF1alpha siRNA and gemcitabine via biocompatible lipid-polymer hybrid nanoparticles for effective treatment of pancreatic cancer. *Biomaterials, 46*, 13–25. https://doi.org/10.1016/j.biomaterials.2014.12.028.

Zhao, D., Wu, J., Li, C., Zhang, H., Li, Z., & Luan, Y. (2017). Precise ratiometric loading of PTX and DOX based on redox-sensitive mixed micelles for cancer therapy. *Colloids and Surfaces B: Biointerfaces, 155*, 51–60. https://doi.org/10.1016/j.colsurfb.2017.03.056.

Zheng, M., Zhong, Z., Zhou, L., Meng, F., Peng, R., & Zhong, Z. (2012). Poly(ethylene oxide) grafted with short polyethylenimine gives DNA polyplexes with superior colloidal stability, low cytotoxicity, and potent in vitro gene transfection under serum conditions. *Biomacromolecules*, *13*(3), 881–888. https://doi.org/10.1021/bm2017965.

Zhou, H., Hou, X., Liu, Y., Zhao, T., Shang, Q., Tang, J., et al. (2016). Superstable magnetic nanoparticles in conjugation with near-infrared dye as a multimodal theranostic platform. *ACS Applied Materials & Interfaces*, *8*(7), 4424–4433. https://doi.org/10.1021/acsami.5b11308.

Zhu, W. J., Yang, S. D., Qu, C. X., Zhu, Q. L., Chen, W. L., Li, F., et al. (2017). Low-density lipoprotein-coupled micelles with reduction and pH dual sensitivity for intelligent co-delivery of paclitaxel and siRNA to breast tumor. *International Journal of Nanomedicine*, *12*, 3375–3393. https://doi.org/10.2147/IJN.S126310.

## FURTHER READING

Bhattacharjee, H., Balabathula, P., & Wood, G. C. (2010). Targeted nanoparticulate drug-delivery systems for treatment of solid tumors: A review. *Therapeutic Delivery*, *1*(5), 713–734.

Greish, K. (2007). Enhanced permeability and retention of macromolecular drugs in solid tumors: A royal gate for targeted anticancer nanomedicines. *Journal of Drug Targeting*, *15*(7–8), 457–464. https://doi.org/10.1080/10611860701539584.

Jiang, X. C., & Gao, J. Q. (2017). Exosomes as novel bio-carriers for gene and drug delivery. *International Journal of Pharmaceutics*, *521*(1–2), 167–175.

Maeda, H., Bharate, G. Y., & Daruwalla, J. (2009). Polymeric drugs for efficient tumor-targeted drug delivery based on EPR-effect. *European Journal of Pharmaceutics and Biopharmaceutics*, *71*(3), 409–419. https://doi.org/10.1016/j.ejpb.2008.11.010.

Singh, R., & Lillard, J. W., Jr. (2009). Nanoparticle-based targeted drug delivery. *Experimental and Molecular Pathology*, *86*(3), 215–223. https://doi.org/10.1016/j.yexmp.2008.12.004.

Yao, J., Feng, J., & Chen, J. (2016). External-stimuli responsive systems for cancer theranostic. *Asian Journal of Pharmaceutical Sciences*, *11*, 585–595.

CHAPTER SIX

# Evaluation of Resveratrol in Cancer Patients and Experimental Models

## Monica A. Valentovic[1]
Toxicology Research Cluster, Joan C. Edward School of Medicine, Huntington, WV, United States
[1]Corresponding author: e-mail address: valentov@marshall.edu

## Contents

1. Introduction — 172
2. Resveratrol and Dietary Sources — 172
3. Resveratrol and Clinical Studies in Cancer Patients — 173
4. RES Mechanisms for Protection — 177
5. Nephroprotection by Resveratrol for Cisplatin — 179
6. Summary — 183
References — 184

## Abstract

Cancer is one of the top three causes of death in the United States. The treatment regimen for controlling cancer includes a number of approaches depending on the classification of the tumor. Treatment may include radiation, surgery, and cancer chemotherapy agents as well as other interventions. Natural products have been identified for centuries to contain active pharmacologic activity and have been a starting point for numerous drugs which are currently on the market. Resveratrol (RES) is a natural product generated in plants in response to environmental stress and growing conditions. RES has been recognized since 1997 to possess anticancer activity. This review discusses the dietary sources of RES and the relative amounts present in the various food sources. A few limited clinical studies have explored RES effects in patients with prostate and colorectal cancer and have suggested some beneficial results. Future studies need to expand the sample size for clinical examination of RES in order to provide a better profile for the potential benefit of RES in cancer patients. This review also describes the potential mechanisms of RES as an antioxidant and in alteration of cell signaling. Another aspect for the role of RES in cancer may be in the interaction with cancer chemotherapy agents. Cisplatin is a cancer chemotherapy agent used for the treatment of bladder, testicular, ovarian, and many other cancers. Cisplatin usage is associated with a high risk of nephrotoxicity. Experimental studies suggest that RES may reduce cisplatin renal toxicity. The proposed mechanisms of protection are reviewed.

## 1. INTRODUCTION

Over 14 million Americans are currently living with some form of cancer based on data collected by the National Cancer Institute. The most commonly diagnosed cancers target the breast, prostate, lung, and colon and rectum. Patients treated for cancer undergo a number of different treatment regimens that include radiation, surgery, and immunotherapy along with the administration of cancer chemotherapy drugs. Treatment with cancer chemotherapy agents is associated with a very high incidence of side effects ranging from nausea, vomiting, alopecia, and bone marrow suppression to more serious side effects of neuropathy, liver failure, and renal toxicity. Extracts of plant material have historically been applied in the treatment of various maladies in humans. Resveratrol (RES) is an endogenous agent present in various food sources such as grapes, blueberries, and peanuts. In the late 1990s, RES was reported to prevent the development of chemical-induced cancers in an experimental animal model. From these initial studies, the activity of RES as an anticancer agent was extensively evaluated in cell culture and in vivo. This review focuses on the sources of RES and the various levels contained in various fruits and foods as well as RES proposed cellular mechanisms for reducing cancer progression. The review provides a synopsis of the results from the relatively sparse clinical studies that have been published involving RES in cancer patients. The review also discusses the potential of RES to reduce the adverse renal effects of a cancer chemotherapy agent.

## 2. RESVERATROL AND DIETARY SOURCES

The chemical *trans*-3,4,5-trihydroxystilbene is known by the more common name of RES. RES is a member of the stilbene family and exists as geometric stereoisomers in the *cis* and *trans* conformation. The *trans* isomer can be converted to the *cis* conformation by photoisomerization. Conversion of the *trans* isomer to the *cis* isomer occurs within minutes in a controlled environment in an acidic environment of pH 4 (Figueiras, Neves-Petersen, & Petersen, 2011). Exposure of grapes to ultraviolet light for 60 s at a wavelength range of 200–280 nm (UVC) increased the levels of RES in common table grapes (Cantos, Espín, & Tomás-Barberán, 2002). The health benefits of RES are focused on the *trans* isomer which exhibits higher anticancer activity; the *trans* isomer will be the focus of this chapter.

RES is a polyphenolic phytoalexin found in the skin of grapes, red wine, chocolate, the root of Japanese Knotweed, peanuts, mulberries, *Vaccinium* plants such as cranberries and blueberries, hops as well as other foods. Japanese knotweed contains the highest amount of RES of dietary plants and is the primary sources for resveratrol supplements. Another plant with worldwide usage and very high RES content are grape products such as the raw fruit, juice, and red wine. Chocolate and cocoa powder contain RES although at a level lower than grape juice or red wine. The RES content is quite heterogeneous between cocoa powder and different types of chocolate. On a per weight basis, RES content is highest in cocoa powder > dark chocolate > milk chocolate (Hurst et al., 2008). A 5-ounce serving of red California wine contains over 800 µg of RES, while a 1.4-ounce serving of dark chocolate contains approximately 14 µg RES. The RES content in beer may be predominantly derived from the hops using in brewing (Olas et al., 2011). The level in beer is variable, but an average 12-ounce serving of beer contains less RES than wine or chocolate (Chiva-Blanch et al., 2011).

RES is generated within the skin of grapes as a defense mechanism in response of the plant following exposure to growing conditions favoring bacterial or fungal growth, ultraviolet radiation, and environmental chemicals (Hasan & Bae, 2017). RES is synthesized in plants from malonyl-CoA by the enzyme stilbene synthase (Hain et al., 1993). Variability in growing conditions and fungal development of powdery mildew will result in variability in RES content of the same grape varieties depending on the geographic location of the grapes. RES levels increase over 4- to 10-fold in some varieties such as Merlot and Pinot Noir grapes as the grape fruit and stems approach ripening and harvest (Geana, Dinca, Ionete, Artem, & Niculescu, 2015).

## 3. RESVERATROL AND CLINICAL STUDIES IN CANCER PATIENTS

The potential beneficial effects of RES have been extensively explored for cardiovascular protection. Moderate ingestion of one to two glasses of red wine was proposed in 1992 to account for the lower incidence of cardiovascular disease in France despite a diet high in saturated fats. This was given the term "French paradox" by Renaud and de Lorgeril who hypothesized that it was red wine ingestion that provided protective factors (Renaud & de Lorgeril, 1992). Renaud and Lorgeril proposed that wine alcohol may impact platelet reactivity to account for the lower incidence of cardiovascular disease in the French population compared to individuals living in the Switzerland, the United States, Scotland, Ireland, China, and Japan. The potential beneficial effects of red wine appear to be dependent on light to moderate ingestion of wine. Heavy alcohol ingestion has a negative impact on health and promotes liver damage, cancer, and cardiovascular disease. Chronic heavy alcohol ingestion increased the incidence of sudden cardiac death and myocardial infarction (Mostofsky, Chahal, Mukamal, Rimm, & Mittleman, 2016; Wannamethee & Shaper, 1992). Red wine contains RES as well as numerous structurally diverse chemicals that potentially contribute toward cardioprotection. Red wine contains not only RES but also anthocyanins, proanthocyanidins, and tannins which vary with the type of grape, growth, and processing practices (Bavaresco, 2003; Waterhouse, 2002). There are numerous reviews regarding the effects of RES on cardiovascular disease including some recent reviews (Mochly-Rosen & Zakhari, 2010; Pagliaro, Santolamazza, Simonelli, & Rubattu, 2015; Sosnowska et al., 2017; Wu & Hsieh, 2011).

There are very limited published studies conducted to examine RES influence to reduce tumor burden in cancer patients. Published clinical studies have reported on the potential effect of RES primarily in colorectal and prostate cancer. RES administration in colorectal cancer patients has shown some beneficial effects when provided prior to surgery. In one of the first published clinical studies, biopsy samples were collected from normal and colon cancer tissue (Nguyen et al., 2009). Individuals subsequently received oral RES tablets (20 or 80 mg/day) or freeze dried California grape powder dissolved in water. Treatment was for 14 consecutive days followed by colon resection. The grape powder was equivalent to ingesting 453 g or higher of raw whole grapes per day. Patel et al. (2010) reported detectable levels of RES in colon tumor and plasma at the end of the 14-day period. Normal tissue showed a decrease in Wnt target gene expression following the 14-day RES treatment. RES and other compounds contained in grape powder did not alter Wnt signaling in cancer tissue, suggesting that RES ingestion may be more critical for normal tissue Wnt signaling.

In another study, 20 patients between the ages of 46 and 83 years were administered RES once daily in the evening for 8 days prior to colon resection. Individuals were given a dose of 500 or 1000 mg/day (Patel et al., 2010). RES was detectable in plasma as well as resected normal and tumor tissue. These findings are significant as they demonstrate which is important since RES has a low bioavailability. An 8-day pretreatment with RES decreased cell proliferation in the tumor tissue. Cell proliferation was compared between normal and cancerous surgical resection tissue by immunohistochemical staining for Ki-67. Ki-67 is a nuclear protein that is an indicator of the frequency of cell division and proliferation. A micronized form of RES was used in a pilot study to determine whether SRT501, a micronized form of RES, could deliver RES to the plasma and tumor tissue in patients with colorectal cancer and hepatic metastasis (Howells et al., 2011). In this study, 20 patients were administered 5 g of SRT501 for 14 days prior to partial hepatectomy. Although this was a limited sample size, caspase-3 cleavage was induced in tumor tissue located in the liver, while normal tissue exhibited no change in caspase cleavage (Howells et al., 2011). The results from the clinical study had a limited number of patients. Further clinical studies need to be conducted to expand to a much larger sample size, but the results suggest that RES may be of some potential benefit in colorectal cancer and specifically in protecting the noncancer colon tissue in cancer patients. These findings are of clinical relevance as colorectal cancer is the second leading cause of death due to cancer in the United States. Interventions that would benefit noncancerous tissue in cancer patients may improve the successful outcome of surgery, radiation, and cancer chemotherapy in patients with colorectal cancer.

A limited number of clinical studies have provided some positive benefits for a select population treated with RES in prostate cancer. An initial pilot study was conducted with 14 males identified with recurrent prostate adenocarcinoma who ingested 500–4000 mg/day of muscadine grape skin powder for 6–29 months (Paller et al., 2015). The average content of RES was 4.4 µg per 500 mg of muscadine grape skin powder. The side effects were predominantly gastrointestinal related to flatulence and soft stools. Muscadine grape skin slowed the doubling time for prostate-specific antigen (PSA) which provides an indirect measure of no metastasis and increased survival. The study did not include a placebo control, and consequently the conclusions regarding clinical benefit are rather limited. A larger cohort study that included 125 males was recently conducted by Paller et al. (2017). In this study, individuals received placebo, 500 or 4000 mg of muscadine grape skin powder daily over a year period administered in 12 cycles of 28 days. Unfortunately, this larger cohort study did show any difference in PSA doubling time between placebo and the different doses of muscadine grape skin powder (Paller et al., 2017). One interesting finding from this study was that men with a single-nucleotide polymorphism for *SOD2* Ala/Ala had a longer doubling time for PSA when comparing the control group. Men with the SOD2 Ala/Ala genotype are associated with a higher risk of prostate cancer (Li et al., 2005). The SOD2 Ala/Ala genotype is associated with a higher activity for manganese superoxide dismutase (MnSOD, SOD2) and may increase oxidative stress in individuals with a low level of antioxidants. Selenium supplementation diminished the risk of prostate cancer, suggesting that the balance of antioxidant/proxidant status was shifted to a ratio more favorable for antioxidant status by selenium. Further studies need to assess whether ingestion of foods rich in antioxidants such as RES can reduce prostate cancer incidence or progression.

The influence of RES on increasing cancer chemotherapy sensitivity or diminishing serious adverse effects is an area with very limited reports. Positive outcomes were reported in one study which suggested that RES may be beneficial in multiple myeloma based on results using cell culture and bone marrow isolates from multiple myeloma patients. In cell culture studies, RES was cytotoxic to multiple myeloma cancer cells by promoting apoptosis (Jazirehi & Bonavida, 2004). RES promoted apoptosis in multiple myeloma cells by diminishing the levels of the antiapoptotic protein Bcl-xL which is often overexpressed in multiple myeloma patients. Independent studies using patient samples and cell culture lines indicated that RES promoted apoptosis in U-266 cells when used at a concentration range of 30–50 µM and enhanced bortezomib-mediated apoptosis (Bhardwaj et al., 2007). The in vitro studies suggested that RES may be beneficial in multiple myeloma. One clinical study has been published that examines whether RES would improve patient response in recurrent multiple myeloma when combined with bortezomib (Popat et al., 2013). The findings were that RES does not appear to be beneficial when combined with the proteasome inhibitor, bortezomib, in patients with relapse multiple myeloma (Popat et al., 2013). A total of 24 patients were treated with a micronized form of RES (SRT501) along with bortezomib. The study was terminated prior to completion due to patient safety concerns. Patients did not show any increase in therapeutic success, but side effects were significant. Patients incurred diarrhea and nausea in over 70% of the participants and over a 20% incidence of renal failure. The cause for the renal impairment was not identified but the mechanism may be attributed to the dehydration mediated by the vomiting and diarrhea. Future approaches may explore compounds structurally similar to RES. One compound is pterostilbene. Pterostilbene is a dimethylated analog or RES and may possess anticancer activity. Pterostilbene, at concentrations of 30–50 µM, exhibited cytotoxicity to H929 multiple myeloma cells by inducing caspase-3 cleavage and apoptosis (Xie et al., 2016). Pterostilbene-mediated cytotoxicity to

H929 cells was associated with an increase in reactive oxygen species (ROS) and loss of mitochondrial membrane potential. Pterostilbene diminished tumor volume in NOD/SCID female mice which were treated daily with 50 mg/kg pterostilbene, but adverse effects to the liver or kidney were not evaluated as part of the in vivo study (Xie et al., 2016).

## 4. RES MECHANISMS FOR PROTECTION

RES anticancer properties have been evaluated for approximately 20 years since the first publication in 1997 (Jang et al., 1997). Jang and colleagues reported that RES reduced the initiation and progression of tumor formation by the carcinogen dimethylbenz(a)anthracene (Jang et al., 1997). RES possesses anticancer activity against numerous cancers including breast, head and neck, bladder, testicular, ovarian, and prostate cancer (Carter, D'Orazio, & Pearson, 2014). Many studies have evaluated the anticancer activity of RES using cancer cell lines to explore the cellular mechanism of protection (Aggarwal et al., 2004; Athar, Back, Kopelovich, Bickers, & Kim, 2009; Jiang et al., 2017; Singh, Ndiaye, & Ahmad, 2015). The potential mechanisms for RES anticancer action that have been suggested include that RES acts as an antioxidant, RES induces apoptosis in cancer cells, and alterations in proteins involved in cell cycle (Ulrich, Wolter, & Steinm, 2005).

Jang et al. (1997) were the first to describe that RES impacted all three stages, initiation, promotion, and growth, of tumor growth. RES reduced cell viability of various cancer cell lines such as breast cancer MCF-7 cells, colon cancer, and A549 lung cancer cells (Takashina et al., 2017). A549 cells treated with 25 µM RES exhibited diminished viability, increased activation of caspase-9, cleavage of caspase-3, and loss of mitochondrial membrane potential (Weng, Yang, Ho, & Yen, 2009). These findings suggest that if sufficient concentrations of RES are delivered to the cancer cell, then RES has a high probability of exerting anticancer activity at least to cultured cancer cells. An in vivo study provided potentially more clinically relevant findings (Fu et al., 2014). RES was protective in vivo in a NOD/SCID mouse xenograft using breast cancer stem cells isolated from MCF-7 cells (Fu et al., 2014). RES treatment of xenograft mice exhibited only one in six mice developing a tumor and increased autophagy in cancer cells. The authors concluded that part of the mechanism for RES protection may be mediated by inhibition of Wnt/β-catenin signaling since overexpression of β-catenin reduced the protective effect of RES.

RES has been recognized to reduce oxidative stress in chemical mixtures, isolated cells, and in vivo. RES may act as a scavenger of ROS as part of its antioxidant activity. A high concentration of 1 mM RES scavenged ROS in vitro to prevent iron-mediated oxidation of linoleic acid (Pignitter et al., 2016). In vivo, an intravenous infusion of RES was sufficient to scavenge ROS in the kidney in anesthetized rats (Gordish & Beierwaltes, 2014), suggesting that RES scavenging may then contribute to increased renal blood flow and a temporary improvement in renal function.

Most studies on the direct antioxidant activity of RES in humans have examined in vitro cell suspension. Oxidative stress mediated by peroxynitrite in human platelet suspensions was diminished in the presence of 50–100 µM RES (Olas, Nowak, Kolodziejczyk, & Wachowicz, 2004). Gresele and associates reported that ingestion by healthy volunteers for 15 days of 300 mL/day of red wine decreased oxidative stress in platelets (Gresele et al., 2008). The group ingesting wine also had diminished NADPH oxidase activity which would imply lower ROS generation. RES was associated with reduced phosphorylation of the p38 mitogen-activated protein kinase (MAPK) which is an indicator of inflammation. In vitro studies also showed that RES increased platelet nitric oxide (NO) levels and NO synthase (Gresele et al., 2008). A follow-up study by Gresele and associates reported that RES stimulated NO synthase and that was not attributed to alcohol as gin alcohol did not produce a similar effect (Gresele et al., 2011).

RES may reduce oxidative stress by increasing the activity of some antioxidant enzymes. The detoxification of ROS relies on a balance of several antioxidant enzymes within the cell. ROS includes several species including superoxide anion, hydrogen peroxide, and hydroxyl radicals. Superoxides are converted to hydrogen peroxide by superoxide dismutases (SODs). MnSOD is localized in the mitochondria, while Cu/ZnSOD is localized in the cytosol. Therefore, depending on the location of the superoxide generation, the ROS will be detoxified by MnSOD or Cu/ZnSOD. The hydrogen peroxide generated will then be detoxified by either catalase or glutathione peroxidase. Glutathione peroxidase requires reduced glutathione with a subsequent formation of glutathione disulfide (GSSG) during hydrogen peroxide detoxification. Glutathione peroxidase can detoxify hydrogen peroxide formed in the mitochondria or the cytosol, provided sufficient levels of reduced glutathione are present. GSSG can be reduced by glutathione reductase in the presence of two molecules of NADPH to generate two molecules of reduced glutathione. As mentioned earlier, catalase is located in the cytosol. Catalase converts hydrogen peroxide to water and oxygen and is present in many cells in the body.

MnSOD dismutates superoxide into hydrogen peroxide which is subsequently converted to water by catalase. RES increased MnSOD, catalase, glutathione peroxidase, and glutathione reductase activity when administered orally to rats simultaneously treated with dimethylhydrazine which is an inducer of colon cancer (Sengottuvelan, Deeptha, & Nalini, 2009). A comparative study of noncancerous HEK293T cells and PC-3 and MCF-7 cells treated with RES showed that the cellular response to RES may vary with the cell type (Khan et al., 2013). Pretreatment of cells with 10–100 µM RES induced cytotoxicity in cancer cells but not in the HEK293T cells. MnSOD activity increased following a 24–72-h exposure to 25 µM RES in the PC-3 and MCF-7 cells but not in the HEK293T cells. Catalase and glutathione peroxidase activity was unchanged by RES concentrations up to 25 µM in HEK293T, MCF-7, and PC-3 cells. Therefore, the lack of increase in activity of catalase and glutathione peroxidase could not compensate for the increased levels of hydrogen peroxide generated in the MCF-7 and PC-3 cells due to increased MnSOD activity. These findings would support the notion that RES mediates cytotoxicity to cancer cells by increasing the intracellular levels of hydrogen peroxide and oxidative stress which would induce cell death. In general, cancer cells have higher levels of oxidative stress (Balliet et al., 2011; Klaunig, Wang, Pu, & Zhou, 2011). Further studies are needed to explore whether RES mediates cell death in cancer cell by altering the balance of antioxidant enzymes. These studies need to also explore many different cell types where the levels of antioxidant enzyme are variable.

## 5. NEPHROPROTECTION BY RESVERATROL FOR CISPLATIN

Another important aspect in the potential usage of RES in the clinics would be as an intervention to reduce the adverse effects of cancer chemotherapy agents. This is a research focus with potentially very high benefits but also increased risk due to the diversity of the human population. Favorable effects may occur only in a fraction of individuals, but this benefit may provide better therapeutic success.

One published study was mentioned earlier, in which bortezomib combined with RES appeared to promote tumor cytotoxicity in a cell culture model (Bhardwaj et al., 2007). A clinical pilot study was conducted in patients with multiple myeloma who had a history of being refractory or a relapse. In this select patient groups RES and bortezomib were not beneficial. The side effects were worsened in patients taking RES and bortezomib (Popat et al., 2013), including renal failure, and the study was stopped prior to completion.

Experimental studies in cell culture and in rodent models have examined the potential of RES to reduce the adverse side effects of the cancer chemotherapy agent cisplatin. There are no published clinical reports examining the combination of RES and cisplatin, and these studies may be warranted in the future if the safety and benefit appear to predict a positive outcome. Cisplatin (*cis*-diamminedichloroplatinum II) is a cancer chemotherapeutic agent used in the treatment of bladder, ovarian, testicular, breast, lung, head and neck, esophageal cancer, and multiple myeloma. Serious adverse effects include neurotoxicity, ototoxicity, and nephrotoxicity. Approximately 33% of patients treated with cisplatin will develop renal dysfunction (Hartmann, Kollmannsberger, Kanz, & Bokemeyer, 1999). The risk for development of renal dysfunction increases with the dose and duration of cisplatin administration. A rate of as high as 60% has been reported in some clinical studies with cisplatin infusions at a high dose in cancer patients (Hall, Diasio, & Goplerud, 1981). Once renal impairment occurs in a patient, cisplatin must be stopped or the dosage reduced. A reduction in dose will potentially lower the success of the cancer chemotherapy treatment regimen. The current clinical intervention used to reduce cisplatin nephrotoxicity is administration of intravenous fluids to hydrate the patient with physiological saline and the osmotic diuretic mannitol. Surprisingly, one study suggested that addition of the osmotic diuretic mannitol did not significantly prevent renal damage relative to hydration (Morgan, Buie, & Savage, 2012). Therefore, other modalities are still needed in the clinics to reduce cisplatin side effects.

Cisplatin toxicity in humans and in rodent experimental models is associated with proximal tubular damage predominantly to the S3 segment as well as the distal tubule of the nephron (Arany & Safirstein, 2003; Pabla & Dong, 2008; Sánchez-González, López-Hernández, López-Novoa, & Morales, 2011). Cisplatin cancer chemotherapeutic mechanism requires entry into the cell followed by aquation of the platinum (Pt) complex. Cisplatin produces inter- and intrastrand cross-linking between the N7 and O6 of guanine nucleotides causing defective DNA which stops DNA synthesis, cell-cycle arrest, and eventual death of the cancer cell (Cepeda et al., 2007; Hambley, 1988). An in-depth review of the chemical interaction between cisplatin and specific sites within DNA is thoroughly reviewed by Cepeda et al. (2007). The cellular mechanism of cisplatin cytotoxicity includes necrosis and high concentrations, activation of caspases and apoptosis, increased oxidative stress, and induction of MAPK and p53 (Cummings & Schnellmann, 2002; Pabla & Dong, 2008). Cisplatin rapidly impairs renal mitochondria and targets the electron transport chain complexes I–IV when added in vitro (Kruidering, Van de Water, de Heer, Mulder, & Nagelkerke, 1997).

Preliminary pilot studies have explored the potential of RES to reduce renal cytotoxicity in noncancerous cells. Glutathione levels were decreased by 30 min with cisplatin in healthy patient platelet suspensions (Olas & Wachowicz, 2004; Olas, Wachowicz, Bald, & Głowacki, 2004). Addition of 25 µg/mL RES with cisplatin partially prevented the decline in cellular thiols. RES appeared to reduce the levels of ROS in platelet suspensions incubated with cisplatin. The reduction in ROS in platelet suspensions incubated with cisplatin had diminished membrane lipid peroxidation (Olas & Wachowicz, 2004). Further studies with cisplatin by this same group reported that RES reduced oxidative stress and DNA damage, as measured using a comet assay, in lymphocytes coincubated with RES and cisplatin (Olas, Wachowicz, Majsterek, & Blasiak, 2005).

Renal cortical slices have been used as an in vitro model to explore renal cortical toxicity of various drugs and other xenobiotics (Berndt, 1976; Burton et al., 1995; Dickman, Sweet, Bonala, Ray, & Wu, 2011; Harmon, Terneus, Kiningham, & Valentovic, 2005; Rankin, Hong, Anestis, Ball, & Valentovic, 2008). Cell culture or isolated cells provide one cell population which may not retain metabolic activity. Renal cortical slices provide a mixed population of cells in the organization retained in the nephron allowing for an interaction of compound generated by one cell type to influence the proximal tubular epithelial cells (Baverel et al., 2013). Our laboratory investigated whether RES would be protective to rat renal cortical slices exposed to cisplatin. The cisplatin level used in the renal cortical slices was comparable to levels detected following in vivo cisplatin injection to rats (Valentovic, Scott, Madan, & Yokel, 1991). The selection of RES concentration for any study is challenging in order to be relevant to the clinical levels. The level of 30 µg/mL RES was based on the peak plasma RES level reported in patients taking 5 g of RES daily for several weeks (Brown et al., 2010). Cisplatin renal cytotoxicity occurred within 90 min to Fischer 344 rat renal cortical slices exposed to 0–150 µg/mL cisplatin (Valentovic et al., 2014). A 30-min pretreatment with 30 µg/mL RES prevented cortical cytotoxicity by cisplatin over a period of 120 min. It was unlikely that RES protection was attributed to diminished uptake of cisplatin as RES was protective when only a 30-min preincubation with RES was followed by a 120-min exposure to only cisplatin. RES reduced oxidative stress mediated by cisplatin exposure as indicated by a reduction in lipid peroxidation, protein carbonylation, and 4-hydroxynonenal protein adduction (Valentovic et al., 2014). A 120-min exposure to cisplatin was sufficient time to decrease MnSOD and glutathione peroxidase enzyme activity, while addition of RES prevented the decline in MnSOD and glutathione peroxidase activity. These findings suggest that RES acted to reduce oxidative stress and maintain cell viability.

In vivo studies have shown the black grape skin extract reduced cisplatin nephrotoxicity in rats (Cetin et al., 2006). The level of RES in the black grape extract was not measured, so specific comparisons for the effect of RES remain unknown. The black grape extract also contains numerous agents that may contribute to the protection observed following a single treatment with cisplatin. It should also be mentioned that in the study by Cetin and colleagues, the grape extract was started 3 days prior to cisplatin administration which may not be feasible in patients. A single treatment with RES was shown by Do Amaral et al. (2008) to be protective for cisplatin nephrotoxicity in male Wistar rats. RES was administered prior to cisplatin but only a single RES treatment. In this study, renal function as measured by creatinine levels, urinary excretion volume, and urinary proteins was improved by RES pretreatment. Examination of renal tissue 2–5 days after cisplatin administration showed that cisplatin mediated an increase in oxidative stress and depletion of glutathione. Cisplatin nephrotoxicity was measured 2–5 days after cisplatin administration because cisplatin is associated with a delay in the manifestations of renal impairment. RES reduced the extent of oxidative stress, suggesting that RES may provide positive benefits in preserving renal function. Do Amaral et al. (2008) also suggested that RES was antiinflammatory as less infiltration of T cells, leukocytes, and macrophages was evident in the renal tissue from rats treated with RES and cisplatin. The infiltration of these inflammatory cells would be expected to generate an inflammatory response and increase localized generation of ROS. In general, mice are more resistant to cisplatin nephrotoxicity compared to rats. However, cisplatin treatment for 7 days induced nephrotoxicity in a C57BL/6 mouse; RES was protective for renal function in C57BL/6 mice treated with cisplatin (Kim et al., 2011). In this study, RES was given orally for 7 days prior to cisplatin administration. Cisplatin was administered at a dose of 10 mg/kg once daily for 7 days by intraperitoneal injection along with oral RES treatment. RES preserved renal function, as indicated by a higher glomerular filtration rate in mice treated with RES + cisplatin when compared to cisplatin alone (Kim et al., 2011). The authors suggested that RES protection was mediated by increased nicotinamide adenine dinucleotide-dependent protein deacetylase 1 (SIRT1) that diminished cisplatin-associated p53 acetylation and activation of apoptotic pathways. Activation of p53 acetylation is thought to be a result of cisplatin damage to DNA causing activation of p53 acetylation that results in the induction of apoptosis through caspase-dependent and -independent pathways (Cummings & Schnellmann, 2002). RES, by stimulation of SIRT1, would diminish acetylation of p53 and induction of cell death. SIRT1 activity declined in cisplatin-treated mouse proximal tubular cells 4–24 h after cisplatin treatment (Kim et al., 2011). This study provided important mechanistic findings that RES may reduce cisplatin nephrotoxicity by increasing SIRT1 expression and promoting deacetylation of p53, thus diminishing apoptosis. In order to examine the mechanism for RES protection further, RES protection was examined in HK-2 cells which are a noncancerous human proximal tubular epithelial cell line. RES protected HK-2 cells from cisplatin cytotoxicity (Jung et al., 2012). More importantly, overexpression of SIRT1 also protected HK-2 cells from cisplatin and activation of p53. RES and SIRT1 overexpression also reduced acetylation of NF-κB p65 subunit (Jung et al., 2012). The prevention of excess acetylation of NF-κB may be an additional mechanism for protection of the kidney from cisplatin nephrotoxicity. However, RES may have multiple mechanisms responsible for its biological effects, and further studies are needed to explore these mechanisms of protection.

## 6. SUMMARY

It is estimated that over 1.7 million new cases of cancer occur each year in the United States. Interventions in diet that may reduce the incidence or severity of cancer are areas of public interest. RES is an endogenous substance synthesized in plants as a response to stress from fungal infections such as powdery mildew and bacterial infections. RES has displayed anticancer properties ranging from inhibiting tumor initiation and tumor progression. The cellular mechanisms for RES are quite varied. Postulated mechanisms for RES include antioxidant, prooxidant, alterations in cellular signaling, and antiinflammatory agent. RES may provide some benefit to patients with certain types of cancer. Clinical studies suggest that patients with colorectal cancer may benefit from RES supplementation prior to surgery. RES may also be beneficial in reducing the adverse effects of some cancer chemotherapy agents. Future studies need to explore RES interactions with cancer chemotherapy agents and optimization of RES delivery to the target tissue.

## REFERENCES

Aggarwal, B. B., Bhardwaj, A., Aggarwal, R. S., Seeram, N. P., Shishodia, S., & Takada, Y. (2004). Role of resveratrol in prevention and therapy of cancer: Preclinical and clinical studies. *Anticancer Research, 24*, 2783–2840.

Arany, I., & Safirstein, R. L. (2003). Cisplatin nephrotoxicity. *Seminars in Nephrology, 23*, 460–464.

Athar, M., Back, J. H., Kopelovich, L., Bickers, D. R., & Kim, A. L. (2009). Multiple molecular targets of resveratrol: Anti-carcinogenic mechanisms. *Archives of Biochemistry and Biophysics, 486*, 95–102.

Balliet, R. M., Capparelli, C., Guido, C., Pestell, T. G., Martinez-Outschoorn, U. E., Lin, Z., et al. (2011). Mitochondrial oxidative stress in cancer-associated fibroblasts drives lactate production, promoting breast cancer tumor growth: Understanding the aging and cancer connection. *Cell Cycle, 10*, 4065–4073.

Bavaresco, L. (2003). Role of viticultural factors on stilbene concentrations of grapes and wine. *Drugs under Experimental and Clinical Research, 29*, 181–187.

Baverel, G., Knouzy, B., Gauthier, C., El Hage, M., Ferrier, B., & Martin, G. (2013). Use of precision-cut renal cortical slices in nephrotoxicity studies. *Xenobiotica, 43*, 54–62.

Berndt, W. O. (1976). Use of the tissue slice technique for evaluation of renal transport processes. *Environmental Health Perspectives, 15*, 73–88.

Bhardwaj, A., Sethi, G., Vadhan-Raj, S., Bueso-Ramos, C., Takada, Y., Gaur, U., et al. (2007). Resveratrol inhibits proliferation, induces apoptosis, and overcomes chemoresistance through down-regulation of STAT3 and nuclear factor-kappaB-regulated antiapoptotic and cell survival gene products in human multiple myeloma cells. *Blood, 109*, 2293–22302.

Brown, V. A., Patel, K. R., Viskaduraki, M., Crowell, J. A., Perloff, M., Booth, T. D., et al. (2010). Repeat Dose Study of the cancer chemopreventive agent resveratrol in healthy volunteers: Safety, pharmacokinetics, and effect on the insulin like growth factor axis. *Cancer Research, 70*, 9003–9011.

Burton, C. A., Hatlelid, K., Divine, K., Carter, D. E., Fernando, Q., Brendel, K., et al. (1995). Glutathione effects on toxicity and uptake of mercuric chloride and sodium arsenite in rabbit renal cortical slices. *Environmental Health Perspectives, 103*(Suppl. 1), 81–84.

Cantos, E., Espín, J. C., & Tomás-Barberán, F. A. (2002). Postharvest stilbene-enrichment of red and white table grape varieties using UV-C irradiation pulses. *Journal of Agricultural and Food Chemistry, 50*, 6322–6329.

Carter, L. G., D'Orazio, J. A., & Pearson, K. J. (2014). Resveratrol and cancer: Focus on in vivo evidence. *Endocrine-Related Cancer, 21*, R209–R225.

Cepeda, V., Fuertes, M. A., Castilla, J., Alonso, C., Quevedo, C., & Pérez, J. M. (2007). Biochemical mechanisms of cisplatin cytotoxicity. *Anti-Cancer Agents in Medicinal Chemistry, 7*, 3–18.

Cetin, R., Devrim, E., Kiliçoğlu, B., Avci, A., Candir, O., & Durak, I. (2006). Cisplatin impairs antioxidant system and causes oxidation in rat kidney tissues: Possible protective roles of natural antioxidant foods. *Journal of Applied Toxicology, 26*, 42–46.

Chiva-Blanch, G., Urpi-Sarda, M., Rotches-Ribalta, M., Zamora-Ros, R., Llorach, R., Lamuela-Raventos, R. M., et al. (2011). Determination of resveratrol and piceid in beer matrices by solid–phase extraction and liquid chromatography-tandem mass spectrometry. *Journal of Chromatography. A*, *1218*, 698–705.

Cummings, B. S., & Schnellmann, R. G. (2002). Cisplatin-induced renal cell apoptosis: Caspase 3-dependent and -independent pathways. *The Journal of Pharmacology and Experimental Therapeutics*, *302*, 8–17.

Dickman, K. G., Sweet, D. H., Bonala, R., Ray, T., & Wu, A. (2011). Physiological and molecular characterization of aristolochic acid transport by the kidney. *The Journal of Pharmacology and Experimental Therapeutics*, *338*, 588–597.

Do Amaral, C. L., Francescato, H. D., Coimbra, T. M., Costa, R. S., Darin, J. D., Antunes, L. M., et al. (2008). Resveratrol attenuates cisplatin-induced nephrotoxicity in rats. *Archives of Toxicology*, *82*, 363–370.

Figueiras, T. S., Neves-Petersen, M. T., & Petersen, S. B. (2011). Activation energy of light induced isomerization of resveratrol. *Journal of Fluorescence*, *21*, 1897–1906.

Fu, Y., Chang, H., Peng, X., Bai, Q., Yi, L., Zhou, Y., et al. (2014). Resveratrol inhibits breast cancer stem-like cells and induces autophagy via suppressing Wnt/β-catenin signaling pathway. *PLoS One*, *9*(7), e102535.

Geana, E. I., Dinca, O. R., Ionete, R. E., Artem, V., & Niculescu, V. C. (2015). Monitoring trans-resveratrol in grape berry skins during ripening and in corresponding wines by HPLC. *Food Technology and Biotechnology*, *53*, 73–80.

Gordish, K. L., & Beierwaltes, W. H. (2014). Resveratrol induces acute endothelium-dependent renal vasodilation mediated through nitric oxide and reactive oxygen species scavenging. *American Journal of Physiology. Renal Physiology*, *306*, F542–F550.

Gresele, P., Cerletti, C., Guglielminim, G., Pignatellim, P., de Gaetanom, G., & Violim, F. (2011). Effects of resveratrol and other wine polyphenols on vascular function: An update. *The Journal of Nutritional Biochemistry*, *22*, 201–211.

Gresele, P., Pignatelli, P., Guglielmini, G., Carnevale, R., Mezzasoma, A. M., Ghiselli, A., et al. (2008). Resveratrol, at concentrations attainable with moderate wine consumption, stimulates human platelet nitric oxide production. *The Journal of Nutrition*, *138*, 1602–1608.

Hain, R., Reif, H. J., Krause, E., Langebartels, R., Kindl, H., Vornam, B., et al. (1993). Disease resistance results from foreign phytoalexin expression in a novel plant. *Nature*, *361*, 153–156.

Hall, D. J., Diasio, R., & Goplerud, D. R. (1981). cis-Platinum in gynecologic cancer. III. Toxicity. *American Journal of Obstetrics and Gynecology*, *141*, 309–312.

Hambley, T. W. (1988). Modelling the interaction of cisplatin with DNA. *Drug Design and Delivery*, *3*, 153–158.

Harmon, R. C., Terneus, M. V., Kiningham, K. K., & Valentovic, M. (2005). Time-dependent effect of p-aminophenol (PAP) toxicity in renal slices and development of oxidative stress. *Toxicology and Applied Pharmacology*, *209*, 86–94.

Hartmann, J. T., Kollmannsberger, C., Kanz, L., & Bokemeyer, C. (1999). Platinum organ toxicity and possible prevention in patients with testicular cancer. *International Journal of Cancer*, *83*, 866–869.

Hasan, M., & Bae, H. (2017). An overview of stress-induced resveratrol synthesis in grapes: Perspectives for resveratrol-enriched grape products. *Molecules*, *22*, 294. https://doi.org/10.3390/molecules22020294.

Howells, L. M., Berry, D. P., Elliott, P. J., Jacobson, E. W., Hoffmann, E., Hegarty, B., et al. (2011). Phase I randomized, double-blind pilot study of micronized resveratrol (SRT501) in patients with hepatic metastases—Safety, pharmacokinetics, and pharmacodynamics. *Cancer Prevention Research (Philadelphia, Pa)*, *9*, 1419–1425.

Hurst, W. J., Glinksi, J. A., Miller, K. B., Apgar, J., Davey, M. H., & Stuart, D. A. (2008). Survey of the *trans*-resveratrol and *trans*-piceid content of cocoa-containing and chocolate products. *Journal of Agricultural and Food Chemistry*, 56, 8374–8378.

Jang, M., Cai, L., Udeani, G. O., Slowing, K. V., Thomas, C. F., Beecher, C. W., et al. (1997). Cancer chemopreventive activity of resveratrol, a natural product derived from grapes. *Science*, 275, 218–220.

Jazirehi, A. R., & Bonavida, B. (2004). Resveratrol modifies the expression of apoptotic regulatory proteins and sensitizes non-Hodgkin's lymphoma and multiple myeloma cell lines to paclitaxel-induced apoptosis. *Molecular Cancer Therapeutics*, 3, 71–84.

Jiang, Z., Chen, K., Cheng, L., Yan, B., Qian, W., Cao, J., et al. (2017). Resveratrol and cancer treatment: Updates. *Annals of the New York Academy of Sciences*, 1403, 59–69.

Jung, Y. J., Lee, J. E., Lee, A. S., Kang, K. P., Lee, S., Park, S. K., et al. (2012). SIRT1 overexpression decreases cisplatin-induced acetylation of NF-κB p65 subunit and cytotoxicity in renal proximal tubule cells. *Biochemical and Biophysical Research Communications*, 419, 206–210.

Khan, M. A., Chen, H. C., Wan, X. X., Tania, M., Xu, A. H., Chen, F. Z., et al. (2013). Regulatory effects of resveratrol on antioxidant enzymes: A mechanism of growth inhibition and apoptosis induction in cancer cells. *Molecules and Cells*, 35, 219–225.

Kim, D. H., Jung, Y. J., Lee, J. E., Lee, A. S., Kang, K. P., Lee, S., et al. (2011). SIRT1 activation by resveratrol ameliorates cisplatin-induced renal injury through deacetylation of p53. *American Journal of Physiology. Renal Physiology*, 301, F427–F435.

Klaunig, J. E., Wang, Z., Pu, X., & Zhou, S. (2011). Oxidative stress and oxidative damage in chemical carcinogenesis. *Toxicology and Applied Pharmacology*, 254, 86–99.

Kruidering, M., Van de Water, B., de Heer, E., Mulder, G. J., & Nagelkerke, J. F. (1997). Cisplatin-induced nephrotoxicity in porcine proximal tubular cells: Mitochondrial dysfunction by inhibition of complexes I to IV of the respiratory chain. *The Journal of Pharmacology and Experimental Therapeutics*, 280, 638–649.

Li, H., Kantoff, P. W., Giovannucci, E., Leitzmann, M. F., Gaziano, J. M., Stampfer, M. J., et al. (2005). 2005. Manganese superoxide dismutase polymorphism, prediagnostic antioxidant status, and risk of clinical significant prostate cancer. *Cancer Research*, 65(6), 2498–2504.

Mochly-Rosen, D., & Zakhari, S. (2010). Focus on: The cardiovascular system: What did we learn from the French (Paradox)? *Alcohol Research & Health*, 33, 76–86.

Morgan, K. P., Buie, L. W., & Savage, S. W. (2012). The role of mannitol as a nephroprotectant in patients receiving cisplatin therapy. *The Annals of Pharmacotherapy*, 46, 276–281.

Mostofsky, E., Chahal, H. S., Mukamal, K. J., Rimm, E. B., & Mittleman, M. A. (2016). Alcohol and immediate risk of cardiovascular events: A systematic review and dose-response meta-analysis. *Circulation*, 133, 979–987.

Nguyen, A. V., Martinez, M., Stamosm, M. J., Moyer, M. P., Planutis, K., Hope, C., et al. (2009). Results of a phase I pilot clinical trial examining the effect of plant-derived resveratrol and grape powder on Wnt pathway target gene expression in colonic mucosa and colon cancer. *Cancer Management and Research*, 1, 25–37.

Olas, K. B., Kolododziejczyk, J., Wachowicz, B., Jedrejek, D., Stochmal, A., & Oleszek, W. (2011). The extract from hop cones (Humulus lupulus) as a modulator of oxidative stress in blood platelets. *Platelets*, 22, 345–352.

Olas, B., Nowak, P., Kolodziejczyk, J., & Wachowicz, B. (2004). The effects of antioxidants on peroxynitrite-induced changes in platelet proteins. *Thrombosis Research*, 113, 399–406.

Olas, B., & Wachowicz, B. (2004). Resveratrol reduces oxidative stress induced by platinum compounds in blood platelets. *General Physiology and Biophysics*, 23, 315–326.

Olas, B., Wachowicz, B., Bald, E., & Głowacki, R. (2004). The protective effects of resveratrol against changes in blood platelet thiols induced by platinum compounds. *Journal of Physiology and Pharmacology, 55*(2), 467–476.

Olas, B., Wachowicz, B., Majsterek, I., & Blasiak, J. (2005). Resveratrol may reduce oxidative stress induced by platinum compounds in human plasma, blood platelets and lymphocytes. *Anticancer Drugs, 16,* 659–665.

Pabla, N., & Dong, Z. (2008). Cisplatin nephrotoxicity: Mechanisms and renoprotective strategies. *Kidney International, 73,* 994–1007.

Pagliaro, B., Santolamazza, C., Simonelli, F., & Rubattu, S. (2015). Phytochemical compounds and protection from cardiovascular diseases: A state of the art. *BioMed Research International, 2015,* 918069.

Paller, C. J., Rudek, M. A., Zhou, X. C., Wagner, W. D., Hudson, T. S., Anders, N., et al. (2015). A phase I study of muscadine grape skin extract in men with biochemically recurrent prostate cancer: Safety, tolerability, and dose determination. *The Prostate, 75,* 1518–1525.

Paller, C. J., Zhou, X. C., Heath, E. I., Taplin, M. E., Mayer, T., Stein, M. N., et al. (2017). Muscadine grape skin extract in men with biochemically recurrent prostate cancer: A randomized, multicenter, placebo-controlled clinical trial. *Clinical Cancer Research.* https://doi.org/10.1158/1078-0432.CCR-17-1100.

Patel, K. R., Brown, V. A., Jones, D., Britton, R. G., Hemingway, D., Miller, A. S., et al. (2010). Clinical pharmacology of resveratrol and its metabolites in colorectal cancer patients. *Cancer Research, 70,* 7392–7399.

Pignitter, M., Schueller, K., Burkon, A., Knorr, V., Esefelder, L., Doberer, D., et al. (2016). Concentration-dependent effects of resveratrol and metabolites on the redox status of human erythrocytes in single-dose studies. *The Journal of Nutritional Biochemistry, 27,* 164–170.

Popat, R., Plesner, T., Davies, F., Cook, G., Cook, M., Elliott, P., et al. (2013). A phase 2 study of SRT501 (resveratrol) with bortezomib for patients with relapsed and or refractory multiple myeloma. *British Journal of Haematology, 160,* 709–722.

Rankin, G. O., Hong, S. K., Anestis, D. K., Ball, J. G., & Valentovic, M. A. (2008). Mechanistic aspects of 4-amino-2,6-dichlorophenol-induced in vitro nephrotoxicity. *Toxicology, 245,* 123–129.

Renaud, S., de Lorgeril, M. (1992). Wine, alcohol, platelets, and the French paradox for coronary heart disease. The Lancet 339,1523–6.

Sánchez-González, P. D., López-Hernández, F. J., López-Novoa, J. M., & Morales, A. I. (2011). An integrative view of the pathophysiological events leading to cisplatin nephrotoxicity. *Critical Reviews in Toxicology, 41,* 803–821.

Sengottuvelan, M., Deeptha, K., & Nalini, N. (2009). Resveratrol ameliorates DNA damage, prooxidant and antioxidant imbalance in 1,2-dimethylhydrazine induced rat colon carcinogenesis. *Chemico-Biological Interactions, 181,* 193–201.

Singh, C. K., Ndiaye, M. A., & Ahmad, N. (2015). Resveratrol and cancer: Challenges for clinical translation. *Biochimica et Biophysica Acta, 1852,* 1178–1185.

Sosnowska, B., Mazidi, M., Penson, P., Gluba-Brzózka, A., Rysz, J., & Banach, M. (2017). The sirtuin family members SIRT1, SIRT3 and SIRT6: Their role in vascular biology and atherogenesis. *Atherosclerosis, 265,* 275–282.

Takashina, M., Inoue, S., Tomihara, K., Tomita, K., Hattori, K., Zhao, Q. L., et al. (2017). Different effect of resveratrol to induction of apoptosis depending on the type of human cancer cells. *International Journal of Oncology, 50,* 787–797.

Ulrich, S., Wolter, F., & Steinm, J. M. (2005). Molecular mechanisms of the chemopreventive effects of resveratrol and its analogs in carcinogenesis. *Molecular Nutrition & Food Research, 49,* 452–461.

Valentovic, M. A., Ball, J. G., Brown, J. M., Terneus, M. V., McQuade, E., Van Meter, S., et al. (2014). Resveratrol attenuates cisplatin renal cortical cytotoxicity by modifying oxidative stress. *Toxicology In Vitro, 28*, 248–257.

Valentovic, M. A., Scott, L. A., Madan, E., & Yokel, R. A. (1991). Renal accumulation and urinary excretion of cisplatin in diabetic rats. *Toxicology, 70*, 151–162.

Wannamethee, G., & Shaper, A. G. (1992). Alcohol and sudden cardiac death. *British Heart Journal, 68*, 443–448.

Waterhouse, A. L. (2002). Wine phenolics. *Annals of the New York Academy of Sciences, 957*, 21–36.

Weng, C. J., Yang, Y. T., Ho, C. T., & Yen, G. C. (2009). Mechanisms of apoptotic effects induced by resveratrol, dibenzoylmethane, and their analogues on human lung carcinoma cells. *Journal of Agricultural and Food Chemistry, 57*, 5235–5243.

Wu, J. M., & Hsieh, T. C. (2011). Resveratrol: A cardioprotective substance. *Annals of the New York Academy of Sciences, 1215*, 16–21.

Xie, B., Xu, Z., Hu, L., Chen, G., Wei, R., Yang, G., et al. (2016). Pterostilbene inhibits human multiple myeloma cells via ERK1/2 and JNK pathway in vitro and in vivo. *International Journal of Molecular Sciences, 17*(11), 1927. pii: E1927.

CPI Antony Rowe
Chippenham, UK
2018-03-14 21:16